Syntheses with
Stable Isotopes

Syntheses with Stable Isotopes

Stable Isotopes

of Carbon, Nitrogen, and Oxygen

Donald G. Ott
Los Alamos Scientific Laboratory
University of California
Los Alamos, New Mexico

A WILEY-INTERSCIENCE PUBLICATION

JOHN WILEY & SONS
New York • Chichester • Brisbane • Toronto

6440-674X √

CHEMISTRY

Library of Congress Cataloging in Publication Data:

Ott, Donald G   1926-
   Syntheses with stable isotopes of carbon,
nitrogen, and oxygen.

   ''A Wiley-Interscience publication.''
   Includes index.
   1.   Chemistry, Organic—Synthesis.
2.   Tracers (Chemistry)   I.   Title.
QD262.O83   547'.2   80-19076
ISBN 0-471-04922-0

Printed in the United States of America

10 9 8 7 6 5 4 3 2 1

Preface

This book is intended to afford a starting place for the chemist confronted with the need to prepare compounds labeled with stable isotopes of carbon, nitrogen, and oxygen. Methods, techniques, ideas, information, and references are presented, which can be used in selecting or devising synthetic schemes that are "best" with respect to the particular purpose and available laboratory capabilities and materials. By studying and comparing methods that other investigators have applied to problems in isotopic labeling, the task of deciding on suitable syntheses for incorporating isotopes into various other compounds can be considerably simplified.

The major portion of the book is devoted to synthetic procedures that have been used for preparation of specific labeled compounds. The descriptions are often given in sufficient detail that they can be applied or modified without necessity for recourse to the original literature. Methods can be compared, feasibility for extensions to other isotope isomers or to related compounds can be assessed, and requirements for apparatus, materials, time, effort, and skills can be evaluated. Additional methods and speculations are presented for a number of other compounds whose syntheses are not given in detail. A few biosynthetic preparations, which afford specific products in good isotopic yield, are described; certain other applications of biological methods are considered briefly.

Arrangement of the procedures into chapters according to functional groups is somewhat arbitrary; that is, not all preparations of carboxylic acids will be found in the chapter dealing with acids and derivatives; certain alcohols appear as components in multistep syntheses in the chapter on hydrocarbons; some compounds could just as well have been placed elsewhere; and so on. Thus it is important to use the index, which lists all labeled compounds in the book, with indication as to whether the citation is for preparation of the material or its use as an intermediate. Preparations of chemically identical, but isotopically different, compounds are grouped together; for example, the preparations of nitrobenzene labeled with either carbon-13, nitrogen-15, or oxygen-18 appear in sequence. The introductory chapter briefly discusses, with leading references, some properties of stable isotopes, analytical methods, and enrichment processes that afford the basic starting materials for syntheses with carbon-12 and carbon-13, nitrogen-14, and nitrogen-15, and oxygen-16, oxygen-17, and oxygen-18. Factors, which dictate that syntheses with isotopes be somewhat different from normal procedures, are presented and certain specialized apparatus and techniques are discussed. These should be given consideration in the course of developing a labeling method that will be the most satisfactory for the particular purpose, laboratory, and investigator. In the introductory paragraphs to the chapters speculations are made concerning some of the newer preparative methods, which are likely to be adapted to isotopic syntheses and probably should be given consideration when new labeling procedures are being sought.

Complete coverage of the literature and inclusion of all syntheses that have been developed for these isotopes have not been attempted, and selection has been necessary to keep the length of the book reasonable. Some elegant schemes have been omitted because important experimental details are lacking in the publications; some because of their similarity to other given procedures; and many because they are essentially the same as the nonisotopic preparations (that is, presence of the label has not affected the experimental procedures). Certain methods presented can obviously be improved. The preparations given represent, in addition to syntheses of specific compounds, *types* of reactions, methods, and techniques that may be applicable to many other, different products. Efficient synthesis of key synthetic intermediates has been considered to be particularly important; obtaining structurally simple, low molecular weight compounds frequently appears to present the biggest stumbling block to building up structurally more complex molecules.

A continuing increase in applications of stable isotopes in all fields of science and technology can be expected, with a concomitant rise in demand for labeled compounds. It is hoped that this book will be helpful, particularly to those uninitiated in isotopic syntheses, in satisfying these needs and in stimulating development of new and improved methods.

I am grateful to Dr. Thomas W. Whaley for many helpful and interesting discussions concerning isotopic synthesis. I am particularly indebted to the late Dr. Wright H. Langham for inspiration and encouragement.

DONALD G. OTT

Los Alamos, New Mexico
November 1980

Contents

Syntheses with
Stable Isotopes

1

Introduction

A tracer or label is any material that accompanies an object of interest, but is not normally present, and thus allows the object to be distinguished from similar ones. Tagging of substances with fluorescent dyes and of molecules with isotopes are well-known examples. Stable (nonradioactive) isotopes, enriched to concentrations higher than those of natural abundance, were becoming established as useful molecular tracers in the late 1930s and early 1940s.[1-4] However, with the greatly increased availability of radionuclides, notably carbon-14 and tritium, in the late 1940s, coupled with the development of simple and efficient analytical methods,[2,5,6] applications of radioactive materials soon far exceeded those of their stable counterparts. Direct tracing of nitrogen and oxygen necessarily remained with the stable isotopes, owing to the inconveniently short half-lives, for most purposes, of the radionuclides.

Although it was obvious that much greater use of carbon-13, nitrogen-15, and oxygen-17 and oxygen-18 could be made in many fields of physical and biological research, this was not happening for two major (probably not independent) reasons—their cost and lack of relatively simple assay methods. In 1969 the Division of Biology and Medicine of the U.S. Atomic Energy Commission initiated a program with a major objective to surmount the cyclical barrier that appeared to be preventing exploitation of the potentials of stable isotopes. That is, the high costs of the materials precluded their widespread use, and the consequent lack of demand for large quantities precluded reductions in costs that would result from larger-scale production. Greatly increased quantities of the stable isotopes of carbon, oxygen, and nitrogen soon became available at very significantly reduced prices,[1,7,8] and were soon being utilized, along with rapid advances in instrumentation that were occurring about the same time (Fourier transform nuclear magnetic resonance spectroscopy, gas chromatography-mass spectrometry with multiple ion detection, and so forth). Applications to various types of studies and developments in isotope production and instrumentation continue to progress steadily and synergistically.[9-14]

Almost all applications of isotopes are dependent on conversion of the materials from isotope production processes into other chemical forms. Utilization of the stable isotopes of carbon, nitrogen, and oxygen usually necessitates their incorpora-

1

tion into organic structures starting from a few basic building blocks, such as carbon monoxide or dioxide, ammonia, and water. There are needs for all types of labeling—some compounds must be specifically labeled in one or more positions, some totally labeled, and some multiply labeled with more than one isotope, both totally and specifically. For some applications high isotopic concentration is required; for others low enrichment is necessary or sufficient. The quantities of compounds needed can range from milligrams to kilograms and may be desired for only single experiments or for repetitive applications. For meeting these varied requirements methods and techniques should be selected and applied that lead to efficient incorporation of the isotopes—ideally, efficient with respect to both materials and effort.

ISOTOPE PROPERTIES, ANALYSIS, AND ENRICHMENT

For a material to be a useful label it must have a detectable property that makes it different from closely related substances. Some nuclear properties that allow essentially chemically identical compounds to be distinguished from each other are given in Table 1. Hydrogen and deuterium are shown for comparison and reference; the two sulfur isotopes illustrate that there are still important materials of great potential usefulness waiting to be developed.

Analytical methods are, of course, based on these differences among the isotopes. Mass spectrometry, nuclear magnetic resonance spectroscopy, vibrational and electronic spectroscopy, density determination, and nuclear reactions are all applicable.[1-5,9-14] The method of choice depends upon the particular isotope, the information desired, the nature and amount of sample, and availability of the

Table 1 Properties of Some Stable Isotopes

Nuclide	Mass	Natural Abundance (%)	Nuclear Spin	NMR Relative Frequency	NMR Relative Sensitivity
^1H	1.0078	99.98	$\frac{1}{2}$	100.00	1.0000
^2H	2.0131	0.02	1	15.35	0.0096
^{12}C	12.0000	98.89	0	−	−
^{13}C	13.0033	1.11	$\frac{1}{2}$	25.14	0.0159
^{14}N	14.0031	99.63	1	7.22	0.0010
^{15}N	15.0001	0.37	$-\frac{1}{2}$	10.13	0.0010
^{16}O	15.9949	99.76	0	−	−
^{17}O	16.9991	0.04	$-\frac{5}{2}$	13.56	0.0291
^{18}O	17.9991	0.20	0	−	−
^{33}S	32.9715	0.74	$\frac{3}{2}$	7.67	0.0023
^{34}S	33.9679	4.22	0	−	−

instrumentation and technical skills. No one method is ideal for every purpose, and what may be considered to be the best method may, in practice, not always be available.

Mass spectrometry is the most generally applicable and widely used method for isotope analysis and can be applied in different ways. It can very accurately determine the overall ratio of isotopes of an element in a sample following conversion to simple molecules, such as carbon dioxide, water, or nitrogen. Or the mass of the molecules themselves and their fragmentation products can be determined—a procedure that often gives, in addition to overall isotopic composition, structural information on how the isotope is distributed within the molecule. When coupled with a gas chromatograph, microprocessor, and computer, an extremely powerful and sensitive technique, both qualitatively and quantitatively, results.

Nuclear magnetic resonance spectroscopy, such an indisputably powerful method for determining many structural and conformational parameters, can also provide simple isotope concentration values through analysis of peak intensities. Multiplets from ^{13}C-^{13}C couplings in multiply labeled compounds (sometimes viewed as a complicating factor) can be used in a variety of ways to obtain various types of information.[15] With a compound doubly labeled at adjacent carbon atoms, such as acetic-$^{13}C_2$ acid, the relative intensities in the split carboxyl resonance give the carbon-13 concentration in the methyl group; and from the intensities of the methyl carbon peaks, the isotope analysis of the carboxyl group is obtained.[1] ^1H-NMR spectra also can often provide carbon-13 analyses (if complex overlapping of peaks does not interfere). Using acetic-$^{13}C_2$ acid again as an example, the signal from the methyl protons is split into a doublet by the carbon-13 in the methyl group and further split by remote coupling with carbon-13 in the carboxyl group. At 100 at. % carbon-13 there would be a doublet of doublets. At lower concentrations the spectrum has nine lines, and by taking appropriate sums and ratios of the peak heights, the isotope concentration at each carbon is readily determined.[16] Even for singly labeled compounds, such as methanol-^{13}C or acetic-2-^{13}C acid, proton spectra can provide carbon-13 analyses, and without necessity for addition of internal standard or a calibration curve.

Many procedures have been developed for isotope ratio determinations in a wide variety of samples encountered in many diverse applications of stable isotopes as tracers.[9-14] Very seldom are such elemental analyses needed in the course of preparation of carbon- or nitrogen-labeled compounds. The isotope concentration in the starting material usually is known and does not change on incorporation into another compound. That is, the carbon-13 concentration in the carboxyl group of an acid prepared by a Grignard reaction is the same as that in the carbon dioxide used for carbonation, and the nitrogen-15 enrichment in an amide can be expected to be the same as that in the starting amine or nitrile. For oxygen labeling, however, with utilization or greater chance for occurrence of isotopic exchange reactions, analysis may be a more necessary component of a synthesis. Ordinarily, when such analyses are required, they are available through the project that initiated the requirement for the synthesis in the first place.

A "black box," analogous to the liquid scintillation counter that has made carbon-14, tritium, and other radioisotopes so easy to use, may someday be developed for stable isotopes. Even without such a device that would provide rapid, automated, isotope analyses, realization of the large amount of information that can be obtained from experiments using present-day instrumentation (which continues to improve at a rapid rate) ensures the increasing utilization of stable isotopes in all fields of science and technology. The stable and radioactive nuclides have their own, unique properties, which, along with many other factors, determine which will be best to use for a particular purpose under particular circumstances.

Tritium and carbon-14, perhaps the best known of all tracers, although radioactive are nevertheless naturally occurring; they have natural abundance concentrations, and their applications in labeled compounds depend on enrichment to higher levels. The stable isotopes of carbon, nitrogen, and oxygen (Table 1) must also be enriched to abnormal concentrations before they can be applied as tracers. At present most of these stable nuclides are separated by taking advantage of small, but significant, differences in vapor pressures of the isotope isomers of certain simple compounds.[7,9,10,17] The two isotopes of carbon are separated by distillation (at cryogenic temperatures) of carbon monoxide. The vapor pressures of $^{13}C^{16}O$ and $^{12}C^{16}O$ are sufficiently different to allow enrichment of the heavy isotope to greater than 90 at. % and carbon-12 to 99.999 at. %. The major impediment to further concentration of the heavy isotope ($^{12}C^{18}O$, which behaves quite similarly to $^{13}C^{16}O$) has recently been ingeniously overcome through a technique of equilibrating, or scrambling, the oxygen isotopes, followed by refractionation to provide $^{13}C^{16}O$ at greater than 99 at. % carbon-13.[17] The extremely long distillation column (205 m, the longest distillation column ever built) resides in a hole in the ground and, with its associated control mechanisms, is a tribute to modern chemical engineering ingenuity.

The five nitrogen and oxygen isotopes are separated similarly by cryogenic distillation of nitric oxide. Owing to more favorable separation parameters, the column does not need to be so long. The product is usually converted to more readily stored and generally usable chemical forms through catalytic hydrogenation to the various isotope isomers of ammonium sulfate and water.

Nothing says that the *major* abundance isotopes (carbon-12, nitrogen-14, oxygen-16) cannot also have unique applications: oxygen-16 in plutonium oxide-powered space auxilliary power supplies in order to lower the by-product neutron background produced by reaction of alpha radiation on other oxygen isotopes; carbon-12 in urea-^{12}C to eliminate the interference produced by natural abundance urea on ^{13}C-NMR spectra of urea-denatured proteins; and nitrogen-14 in field-scale studies of fertilizers. Applications of such carbon-12 compounds and of ^{12}C-enriched (^{13}C-depleted) solvents in NMR studies are obvious and will, sooner or later, constitute significant uses for this isotope. Other applications for the light isotopes, such as diminishing the carbon-13, nitrogen-15, and oxygen-18 (natural) backgrounds in microbiological tracer experiments by first growing up cultures on carbon-12, nitrogen-14, and oxygen-16 substrates, preliminary to applying the heavy isotope tracers, will undoubtedly become more commonplace.

GENERAL SYNTHETIC CONSIDERATIONS, APPARATUS, AND TECHNIQUES

Syntheses of isotopically labeled compounds usually proceed by way of common organic and biochemical reactions and techniques that are well-known and understood. However, there are certain considerations that often require somewhat different approaches to devising and carrying out such preparations than would normally be used with nonisotopic materials. Indeed, many compounds that must be labeled are almost never synthesized otherwise on a laboratory scale (except perhaps as educational exercises), because they are such common reagents, readily available commercially. A compound can be labeled in a number of different ways—at one or more different molecular positions or with different isotopes, either singly or in combination—and a route developed for a particular isotope isomer may not be the choice for another. Because of this wide variety of possible labeled intermediates, even of relatively simple structures, they are not routinely available "off the shelf." It is frequently necessary to start an isotopic preparation at a quite fundamental level using very basic starting materials, such as water, ammonia, and carbon monoxide, with which many chemists have had little experience in applying as yield-determining reagents.

The cost of enriched isotopes is, of course, a good reason for wanting to utilize high-yield reactions and usually warrants expenditure of research and development time to obtain them. It is an important consideration, but should be kept in proper perspective with other aspects of the project. Prices are often presented in such terms as dollars per gram or liter; a better evaluation can usually be made if comparisons are on a molar basis. The cost of basic isotopic starting materials is of the order of 10^3/mol,[7c] which is in the price range of a number of commercially available, nonisotopic chemicals and reagents (many of which are quite commonly used).

Efficient, high-yield procedures are always desirable, and the development of each new synthesis carries the additional benefit of simplification of future problems. Particularly valuable are proven procedures for preparing key synthetic intermediates and basic compounds, which can then be used in introducing specific labels into a variety of more complex molecules. From economic considerations, however, the time and effort expended to develop or improve a scheme to provide a high yield based on isotope should be balanced against the cost of the starting materials. When large quantities of product are needed, or the method is to be used repetitively, the work necessary to improve the yield or the ease of carrying out a procedure is usually justifiable. But it is sometimes all too easy to spend far too much time, from an economics point of view, on making minor, albeit esthetically pleasing, improvements.[18]

When devising a scheme for synthesis of a labeled compound, the normal literature search should be carried out. If a method has already been developed for labeling the desired material, then the problem is simplified considerably, even if modifications are needed for the particular situation. Whether a procedure is found or not, the various specialized publications[2,3,5,19-21] dealing with the subjects of labeling and labeled compounds are consulted. Especially valuable is the extensive treatment by Murray and Williams,[5] not only for reaction schemes, but for apparatus and techniques as well. Preparations of labeled compounds are to be found in many journals, although the *Journal of Labelled Compounds and Radiopharmaceuticals* is, naturally, by far the richest source. Ideas, or even directly applicable procedures, can often be obtained by examining methods for closely related compounds; for example, if 2-amino-pentanoic-^{15}N acid is desired, methods used for glycine-^{15}N and alanine-^{15}N (as well as the carbon-13 isomers) should be considered.

A preparation that has been used for one particular nuclide is often readily transposed to another. For example, a synthesis of deuterated or tritiated benzaldehyde by reduction of benzoyl chloride can very likely be applied directly to the carbon-13 isomers; and in the incorporation of a labeled pyrimidine into a nucleoside it makes no difference whether the isotope is carbon-13 or nitrogen-15. Sometimes, however, an elegant scheme developed for labeling with a radionuclide cannot be effectively transferred to a stable isotope because of the larger scale usually required or the necessity to include an unlabeled carrier. In comparing methods for the carbon isotopes, it should be remembered that carbon-^{13}C monoxide is a readily available (primary) starting material (in contrast to carbon-^{14}C monoxide). As applications develop in the future for chlorine-35 and chlorine-37 and sulfur-33 and sulfur-34, preparations of the required labeled compounds will undoubtedly rely heavily on procedures used[3,5] for incorporation of the radioactive isotopes chlorine-36 and sulfur-35.

When previously developed isotopic syntheses that can be used or adapted are lacking or felt to be unsuitable for the purpose, a new method must be devised. A considerable number of reactions that are standard practice can be applied to labeling compounds, while keeping in mind conservation of the isotope and the nature of available starting materials. As for any new synthesis development there are many publications dealing with the subject that are very helpful in the search for or confirmation of ideas for a suitable approach. Among these are the treatments of Buehler and Pearson,[22] Fieser and Fieser,[23] Harrison and Harrison,[24] *Organic Syntheses*,[25] *The Merck Index*,[26] and various laboratory texts. Experienced synthetic chemists progress through the design process almost automatically; however, those less familiar with the operation should not consider it such a specialized field that it is outside their capabilities. Many excellent isotopic preparations have been devised by investigators who do not claim expertise in synthetic organic chemistry.

The majority of labeled compounds are prepared by organic synthesis, although biosynthesis[27] is often the method of choice, if not the only *practical* method, for obtaining certain materials—particularly those with complex structures and with numerous chiral centers. The organic chemist should not fail to consider such possibilities, even if he is unfamiliar with microbiological and biochemical techniques. They are often within his normal realm of capabilities and laboratory apparatus. Procedures that afford specific compounds in high isotopic yield are particularly useful, such as the photosynthetic production of glucose-$^{13}C_6$ by excised green leaves of plants, or of galactosylglycerol-$^{13}C_9$ by thalli of marine algae, from carbon-^{13}C dioxide.[28] Specific labeling of particular entities and molecular sites in complicated biomolecules is possible through a combination of organic and biosynthesis; for example, incorporation of L-histidine-2-^{13}C into hemoglobin for NMR studies.[29] Combined organic-biosynthesis will likely emerge as the method of choice for certain other types of specifically labeled compounds, particularly amino acids. A number of microbiological mutants are known or are being developed[30] that could be applied advantageously; for example, tryptophan could be labeled in various positions (multiple if desired) by using the appropriately labeled indole and/or serine as substrates for a particular indole- and/or serine-requiring bacterial strain. Such combined syntheses appear to offer great potential for supplying the demand for this most important class of labeled compounds. On the other hand, production of amino acids (uniformly or generally labeled) through photosynthetic incorporation of isotopes by algae or other organisms, followed by hydrolysis of the protein and subsequent separation of the components, does not appear to be either an efficient or an economical procedure for most purposes. It may be useful, however, when only small amounts of a number of amino acids are required, such as for mass spectral internal standards.

Whatever type of synthetic route may be used, frequently the isolation and purification steps are the most difficult parts and can have a greater effect on yield and effort required than does carrying out the chemical reaction. Often the purification method is determined by the scale of the reaction. For example, large amounts of material are seldom amenable to chromatographic techniques, and distillation or crystallization can be used with little mechanical loss; for small quantities the reverse is usually true. Purification problems are often simplified by knowing fairly specifically what the application of the product will be. It may be a waste of time, effort, and material to isolate a pure, crystalline substance from aqueous solution, if the next step is going to involve dissolving it in water anyway. Also, the requirements for the product should be considered in determining what is the most practical chemical form in which to isolate the material. For example, an acid or a base may be more difficult to isolate than a salt, which may serve the purpose just as well.

The decision as to which of the possible synthesis schemes and isolation and purification procedures are best for the purpose depends not only on the yield to be expected, but also on the amount of effort that will be expended in conjunction with the particular equipment, reagents, capabilities, experience, and personnel that are available in the laboratory in which the synthesis will be conducted. A procedure that is most advantageous for one laboratory may not be for another.[18] The chosen method must, of course, meet the requirements imposed for the product: the particular isotope(s) and concentration(s), the molecular site(s) to be labeled, the quantity and purity, possible future and additional uses, and the allowed time period. It is highly advisable that the specifications be as realistic as possible, and that it be determined whether certain deviations might be allowable if unforeseen difficulties arise.

The presence of an isotope in a reaction scheme does not necessarily mean that normal laboratory techniques and procedures must be supplanted by operations that are new and different from those normally employed in synthetic chemistry. However, the frequent necessity to utilize low-molecular weight labeled intermediates often requires applications of procedures that are not usually applied in the course of most, conventional, organic preparations. Ordinarily when usage of those materials in nonisotopic form is required, they are readily available as low-cost reagents and often employed in excess without any need for efficient utilization. The basic isotopic starting materials and many key synthetic intermediates (carbon monoxide and dioxide, ammonia, oxygen, methane, acetylene, water, hydrogen cyanide, various alcohols, halides, amines, and many others) have vapor pressures high enough that handling them as gases is either required, convenient, or efficient. Owing to their cost or availability, they cannot be used in excess or wasted, as might be the most efficient way for application of their unlabeled analogs. Conducting reactions in closed systems is therefore a useful procedure. For example, carbonation of a Grignard reagent with carbon-^{13}C dioxide can be carried out by replacing the usual reflux condenser with a vacuum gauge, attaching the carbon dioxide source (a closed generator or cylinder), cooling the mixture to the desired temperature, and evacuating to the vapor pressure of the solvent. The carbon dioxide is admitted at a rate determined from the pressure gauge.

Closed systems can be assemblies for such particular applications, or they can be constructed in ways that serve a number of different purposes. In a laboratory in which varied types of preparations are carried out, a manifold assembled from stainless steel components can be very useful. A metal system has the advantage over one of glass in that it can operate under relatively high pressures as well as vacuum. It can be constructed from tubing and pipe, with appropriate fittings and valves (Swagelok, Cajun, Nupro, Whitey, and others), pressure and vacuum gauges, inlets, outlets, reservoirs, and so forth. Most chemists are more familiar with assembling glass apparatus; however, with a little thought and practice, proficiency with the steel components can soon be attained, along with a preference for this type of system for many applications.

Reaction vessels can be conventional glassware (with appropriate adapters for attachment to the manifold) for low-pressure reactions. For high-pressure procedures autoclaves of various types can be used; however, stainless steel gas sampling cylinders, which come in many sizes (40-, 75-, 150-, 300-, and 1000-cm^3 volumes and others), are very well suited as reaction vessels for many purposes, as well as economical. Magnetic stirring is effected by the introduction of a magnetic bar (Teflon-coated) into the cylinder before the valve and other fittings are attached. The stirring motor is often more effective when rotated 90° from its normally used position and placed along the vertical (or tilted) cylinder wall, which causes the stirring bar to flip up and down and provide quite effective agitation of the contents. A shaker can also be used if the cylinder does not need to remain attached to a manifold or other fixed components. Heating can be by electrical tape or oil bath. Solid reactants are introduced before attaching the valve; liquids can also be added in this manner or by drawing them into the assembled, evacuated cylinder. Solid and high-boiling liquid products can be removed after the reaction by detaching the fittings.

The cylinders are also useful in transfers of specific amounts of gaseous or low-boiling materials. The manifold can be fitted with cylinders of appropriate volumes as reservoirs, and through application of pressure-volume calculations a determined quantity of material can be cryogenically transferred from a storage vessel to a smaller cylinder or reaction mixture. With gases such as ammonia or carbon dioxide that liquefy at relatively low pressures, the nonlinearity of the pressure-volume (pressure-quantity) relationship should be considered; even so, in practice, little correction is required over a considerable range, and calibration curves can be made by weighing transferred amounts of material that the apparatus contained at certain pressures. Prior to transfer the system is evacuated with a good mechanical pump; extremely high vacuum is not required, and a diffusion pump is unnecessary. Although carbon monoxide has a vapor pressure of almost one atmosphere at liquid nitrogen temperature, it can nevertheless be transferred very well cryogenically without necessity for complicated compressor systems or Toepler pumping, which is inconvenient at the scale of most syntheses with stable isotopes. In this case the volume of the system between the storage cylinder and reaction vessel should be kept as small as possible; the amount transferred can be determined from the volume and decrease in pressure of the storage cylinder (confirmed by weighing when possible).

Certain of the basic isotopic starting materials are available in different chemical forms. Whether to start with carbon-^{13}C dioxide or barium carbonate-^{13}C, or with ammonium-^{15}N sulfate or ammonia-^{15}N, depends on a number of considerations similar to those involved in deciding what constitutes the best synthetic scheme. The cost (for contained isotope)[7c] is the same or similar for the solid forms as for the gases; however, packaging costs are greater for the latter. Thus the actual costs for procuring small amounts of isotopes will be less for the solids. Generation of ammonia, particularly if it must be anhydrous, from an ammonium salt, or of carbon dioxide from a carbonate, is conceptually and technically a straightforward operation; but it does require effort, time, and equipment—often more than does the succeeding organic reaction. With some reactions the salts can or should be used; where there is a choice the decision will depend on the particular situation. Some investigators feel there is less chance for inadvertent loss of a solid than of a gas. With carbon monoxide such an alternative is not available; but if storage as a solid was desirable for particular applications, sodium formate might be given consideration.

ISOTOPIC STRUCTURAL DESIGNATIONS AND NOMENCLATURE

Conventions for designation of the locations of molecular positions that are isotopically enriched over natural abundance in structural formulas have become fairly standardized and unambiguous. Thus in the structure

$$PhCO*CH_3$$

it is obvious that the carbon atom of the methyl group of acetophenone has greater than normal abundance of one of the carbon isotopes. More specifically,

$$PhCO^{13}CH_3$$

indicates that the enriched isotope is carbon-13.

From certain reactions, such as

$$PhCOCH_3 + *CH_3MgX \longrightarrow Ph-\overset{\displaystyle OH}{\underset{\displaystyle *CH_3}{\overset{|}{\underset{|}{C}}}}-CH_3$$

$$\longrightarrow Ph-\underset{\displaystyle *CH_3}{\overset{|}{C}}=CH_2 + Ph-\underset{\displaystyle CH_3}{\overset{|}{C}}=*CH_2 \tag{1}$$

the natural inclination is to avoid repeating the chemically identical structures and show the mixture of products as

$$Ph-\underset{\displaystyle *CH_3}{\overset{|}{C}}=*CH_2$$

But the same representation results from a reaction that gives the truly double-labeled compound:

$$PhCO*CH_3 + *CH_3MgX \longrightarrow Ph-\overset{\displaystyle OH}{\underset{\displaystyle *CH_3}{\overset{|}{\underset{|}{C}}}}-*CH_3$$

$$\longrightarrow Ph-\underset{\displaystyle *CH_3}{\overset{|}{C}}=*CH_2 \tag{2}$$

The products of the two reactions are distinct isotope isomers, and this should be made apparent in their formulas. The difficulty arises because of the desire for a shorthand notation, even though this is not expected with nonisotopic isomers, such as

$$Ph-\underset{\underset{CH_3}{|}}{C}=CH_2 \text{ and } Ph-CH=CH-CH_3$$

In this book the following convention has been adopted for products that are mixtures of isotope isomers:

$$Ph-\underset{\underset{*CH_3}{|}}{\overset{\overset{OH}{|}}{C}}-CH_3 \longrightarrow Ph-\underset{\underset{^xCH_3}{|}}{C}=^xCH_2 \qquad (3)$$

and

$$(4)$$

Another shorthand scheme is used when a procedure is applied to preparations of more than one isotope isomer, such as

$$*CO + 2^\bullet NH_3 \xrightarrow[\text{MeOH}]{\text{S}} H_2{}^\bullet N*CO^\bullet NH_2 \qquad (5)$$

This indicates the method has been used to synthesize various isotope isomers of urea. A number of preparations are obviously applicable to other isotope isomers as well, but they are not shown in this way unless they have actually been applied.

Assigning a name to a labeled compound, which unambiguously describes its chemical and isotopic composition, is not so straightforward. Several systems have evolved, but none is entirely satisfactory, and all require elaborate rules. The difficulties arise, for the most part, because of the desire to give a *single* name to a *mixture* of compounds. For example, equation 4 gives a mixture of two isotope isomers, which is conveniently called 4-nitroaniline-1,4-$^{13}C_1$. However, in an analogous nonisotopic reaction

the mixture of two isomeric products would simply be considered a mixture of

2-amino-5-nitrobenzoic acid and 5-amino-2-nitrobenzoic acid, without attempting a shorthand combination of two names into one. Nevertheless, with isotopic nomenclature, abbreviation systems are well established, convenient, useful, and additionally warranted in that there is less diversion from primary focus on the chemical structure.

The first requirement that must be met in naming a labeled compound is to use an *acceptable* chemical name. The book *Nomenclature of Organic Compounds*[31] is an excellent treatment of the subject with a large number of helpful examples and explanations, as well as of the interrelationships of various systems—Chemical Abstracts Service, International Union of Pure and Applied Chemistry (IUPAC), and common or trivial names. Several systems are in use for incorporating the necessary information about the kind(s) and location(s) of the isotope(s). *Chemical Abstracts* employs an extension of the principles proposed by Boughton and detailed in its *Index Guide*. An extensive treatment of isotopic organic nomenclature, consistent with *Chemical Abstracts*, is given by Murray and Williams.[5] It deals with situations that cannot be handled by the *Chemical Abstracts* (or other) system; however, many of the features, which allow naming of quite complicated structures, have not been widely adopted, probably because considerable familiarity with the rules is required. Recently the IUPAC published recommendations for nomenclature of isotopically modified compounds.[32] Again, proper application necessitates thorough familiarity with the rules and definitions, many of which may seem strange, especially to those accustomed to Boughton-type systems. Various journals have their own requirements as to how isotopic designations are to be made; sometimes these allow interpretable names for only the simpler compounds, and structural formulas are needed to be certain of the labeled sites.

The isotopic nomenclature used in this book is essentially that of Murray and Williams[5] and *Chemical Abstracts*, in which the isotopic designations ordinarily appear as suffixes. The following examples do not include all possibilities or present many of the rules (which for the most part are needed only for relatively rare cases anyway), but illustrate some common usages.

Equation 1 and 3	2-Phenylpropene-1,3-$^{13}C_1$
Equation 2	2-Phenylpropene-1,3-$^{13}C_2$
Equation 4	4-Nitroaniline-1,4-$^{13}C_1$
Equation 5	Urea-^{12}C, -^{13}C, -$^{14}N_2$, -$^{15}N_2$, -^{13}C-$^{15}N_2$, ...

$$CH_3CH_2OH \xrightarrow{\ ^{13}CO\ } CH_3CH_2{}^{13}COOH \qquad \text{Propionic-1-}^{13}C \text{ Acid}$$

$$^{13}CH_3CH_2OH \xrightarrow{\ CO\ } \begin{array}{c} ^{13}CH_3CH_2COOH \\ + \\ CH_3{}^{13}CH_2COOH \end{array} \qquad \text{Propionic-2,3-}^{13}C_1 \text{ Acid}$$

$$^{13}CH_3{}^{13}CH_2OH \xrightarrow{\ CO\ } {}^{13}CH_3{}^{13}CH_2COOH \qquad \text{Propionic-2,3-}^{13}C_2 \text{ Acid}$$

$$^{13}CH_3{}^{13}CH_2OH \xrightarrow{\ ^{13}CO\ } {}^{13}CH_3{}^{13}CH_2{}^{13}COOH \qquad \text{Propionic-}^{13}C_3 \text{ Acid}$$

When the position of label is indicated by its inclusion in a substituting group, as in 1-(methyl-^{13}C)cyclohexene, the group is considered complex [as is, for example, (chloromethyl) in conventional nomenclature], and parentheses are used. Some compounds lack normal locants for certain positions, which are indicated by using the function as a locant, such as in 3-cinnolinecarboxylic-*carboxy*,3,4-$^{13}C_3$ acid. When no confusion will result, the "odd" locant is omitted and the label is understood as being at the functional group of the parent compound, as 3,4-(methylene-dioxy)benzoic-^{13}C acid. Indefinite isotopic designation, as a prefix, is used when the position is not known or specified, or if no locant exists: $^{13}C_1$-maleic hydrazide for 2,3-dihydro-3,6-pyridazinedione-3-^{13}C; or ^{15}N-lysine as a title for a procedure that can give either lysine-N^2-^{15}N, lysine-N^6-^{15}N, or lysine-$^{15}N_2$.

Both nomenclature and structure are independent of the degree of isotopic enrichment. The same name and same structural designations are used whether the isotope concentration at the labeled position is high or low. For example, the product from the reaction

$$BrCH_2CH_2Br + 2K^{13}CN \xrightarrow[\text{2. } H_2SO_4]{\text{1. EtOH}} HOO^{13}CCH_2CH_2{}^{13}COOH$$

is succinic-1,4-$^{13}C_2$ acid, even though from potassium cyanide-^{13}C at 10 at. % carbon-13 only 1% of the product molecules would actually be doubly labeled (there are instances in the literature that, merely because of low isotope concentration, have given the singly labeled names, such as succinic-1-^{13}C acid). In the experimental procedures given in the following chapters isotopic concentrations are indicated only when they are important to the course of the reactions; except for exchange or dilution processes, synthetic methods are generally independent of the degree of enrichment of the isotope.

When the enriched isotope happens to be one of major natural abundance (carbon-12, nitrogen-14, oxygen-16), the nomenclature system remains the same; that is, for

$$^{12}CH_3{}^{12}COOH$$

the name is acetic-$^{12}C_2$ acid, which shows that the carbons are enriched in carbon-12 over natural abundance. There is no need to invoke such names as acetic-*def*-^{13}C to indicate a deficiency from normal abundance of carbon-13. Also, designation of the major abundance isotope should not be used to indicate natural abundance materials. Normal isotopic concentration is assumed when a mass number is not specified; if emphasis is felt to be necessary, "n" or "na" can be used.

Nomenclature is perhaps somewhat like grammar—it may not have to be proper to convey the intended meaning, but there is less chance for misunderstanding when it is.

References

[1]N. A. Matwiyoff and D. G. Ott, "Stable Isotope Tracers in the Life Sciences and Medicine," *Science*, **181**, 1125-1133 (1973).

[2]M. Calvin, C. Heidelberger, J. C. Reid, B. M. Tolbert, and P. M. Yankwich, *Isotopic Carbon*, Wiley, New York, 1949.

[3]R. H. Herber, Ed., *Inorganic Isotopic Syntheses*, W. A. Benjamin, New York, 1962.

[4]P. D. Klein, D. L. Hachey, M. J. Kreek, and D. A. Schoeller, "Stable Isotopes: Essential Tools in Biological and Medical Research," in *Stable Isotopes*, T. A. Baillie, Ed., University Park Press, Baltimore, 1977, pp. 3-14.

[5]A. Murray, III, and D. L. Williams, *Organic Syntheses with Isotopes*, Interscience, New York, 1958.

[6]D. G. Ott, "History of Liquid Scintillator Development at Los Alamos," in *Proceedings of the International Conference on Liquid Scintillation Counting, San Francisco, CA, 1979*, C. T. Peng and D. L. Horrocks, Eds., Academic, New York, 1980, pp. 1-10.

[7](a) D. E. Armstrong, A. C. Briesmeister, B. B. McInteer, and R. M. Potter, *A Carbon-13 Production Plant Using Carbon Monoxide Distillation*, Los Alamos Scientific Laboratory Report LA-4391 (1970); (b) R. E. Schreiber, Ed., *ICONS at LASL*, Los Alamos Scientific Laboratory Report LA-4759-MS (1971); (c) Stable isotopes produced by the separation facility at Los Alamos Scientific Laboratory in excess of requirements of the Department of Energy are transferred to the Stable Isotope Sales outlet at the Mound Facility, Miamisburg, Ohio, for distribution (at cost) to the general public; (d) Other suppliers of basic starting materials and producers of labeled compounds may be found in *Guide to Scientific Instruments*, American Association for Advancement of Science (current edition).

[8]C. F. Eck, "Carbon-13, the Growing Isotope," *Res./Dev.*, **Aug. 1973**, 32-38.

[9]P. D. Klein and L. J. Roth, Eds., *Proceedings of a Seminar on the Use of Stable Isotopes in Clinical Pharmacology*, Chicago, IL, U. S. Atomic Energy Commission CONF-711115, 1971.

[10](a) P. D. Klein and S. V. Peterson, Eds., *Proceedings of the First International Conference on Stable Isotopes in Chemistry, Biology, and Medicine*, Argonne, IL, U. S. Atomic Energy Commission CONF-730525, 1973; (b) E. R. Klein and P. D. Klein, Eds., *Proceedings of the Second International Conference on Stable Isotopes*, Oak Brook, IL, U. S. Energy Research and Development Administration CONF-751027, 1975; (c) E. R. Klein and P. D. Klein, *Stable Isotopes, Proceedings of the Third International Conference*, Oak Brook, IL, Academic, New York, 1979.

[11]International Atomic Energy Agency, *Proceedings of a Symposium on New Developments in Radiopharmaceuticals and Labelled Compounds*, Copenhagen, Denmark, IAEA, Vienna, 1973; (b) International Atomic Energy Agency, *Proceedings of a Symposium on Isotope Ratios as Pollutant Source and Behavior Indicators*, IAEA, Vienna, 1975.

[12]T. A. Baillie, Ed., *Stable Isotopes*, University Park Press, Baltimore, 1977.

[13]E. R. Klein and P. D. Klein, "A Selected Bibliography on Biomedical and Environmental Applications of Stable Isotopes. I. Deuterium 1971-1976; II. ^{13}C 1971- 1976; III. ^{15}N 1971-1976; IV. ^{17}O, ^{18}O and ^{34}S 1971-1976," *Biomed. Mass Spectrom.*, **5**, 91-111; 321-330; 373-379; 425-432 (1978).

[14]KOR Isotopes, *Research Bibliography. Stable Isotopes in Diagnosis: Research and Applications*, Cambridge, MA, 1978.

[15]R. E. London, V. H. Kollman, and N. A. Matwiyoff, "Biosynthetic and Biophysical Information from ^{13}C-^{13}C Multiplets by ^{13}C Nuclear Magnetic Resonance," in reference 10b, pp. 470-484.

[16]D. G. Ott and V. N. Kerr, "Syntheses with Stable Isotopes: Acetic-1-^{13}C, -2-^{13}C, -$^{13}C_2$, and -$^{12}C_2$ Acids," *J. Labelled Compds.*, **12**, 119 (1976).

[17]B. B. McInteer, "Isotope Separation by Distillation: Design of a Carbon-13 Plant," in *Separation Science Technology*, A. P. Malinauskas, Ed., Dekker, New York (in press).

[18]D. G. Ott, "One-Carbon ^{13}C-Labeled Synthetic Intermediates. Comparison and Evaluation of Preparative Methods," in reference 10c, pp. 11-17.

[19] A. F. Thomas, *Deuterium Labeling in Organic Chemistry*, Appleton-Century-Crofts, New York, 1971.

[20] E. Buncel and C. C. Lee, Eds., *Isotopes in Organic Chemistry*. Vol. 1, *Isotopes in Molecular Rearrangements*; Vol. 2, *Isotopes in Hydrogen Transfer Processes*; Vol. 3, *Carbon-13 in Organic Chemistry*; Vol. 4, *Tritium in Organic Chemistry*, Elsevier, New York, 1975, 1976, 1977, and 1979.

[21] E. A. Evans, *Radiotracer Techniques and Applications*, Dekker, New York, 1977.

[22] C. A. Buehler and D. E. Pearson, *Survey of Organic Syntheses*, Vols. 1 and 2, Wiley-Interscience, New York, 1970 and 1977.

[23] L. F. Fieser and M. Fieser, *Reagents for Organic Synthesis*, Wiley, New York, 1979; M. Fieser and L. F. Fieser, *Reagents for Organic Synthesis*, Vols. 2-7, Wiley-Interscience, New York, 1969, 1972, 1974, 1975, 1977, and 1979.

[24] I. T. Harrison and S. Harrison, *Compendium of Organic Synthetic Methods*, Wiley-Interscience, New York, 1971.

[25] *Org. Syn.*, Coll. Vols. I, II, III, IV, and V (1941, 1943, 1955, 1963, 1973) and subsequent annual issues.

[26] M. Windholz, Ed., *The Merck Index*, 9th ed., Merck and Co., Inc., Rahway, NJ, 1976.

[27] H. W. Whitlock, Jr., "Enzymatic Versus Chemical Synthesis of Molecules Labeled with Heavy Isotopes," in *Techniques of Chemistry. Applications of Biochemical Systems in Organic Chemistry*, Vol. X, Part 2, Ch. XI, J. B. Jones, C. J. Sih, and D. Perlman, Eds. (A. Weissberger, Series Ed.), Wiley-Interscience, New York, 1976.

[28] (a) V. H. Kollman, J. L. Hanners, J. Y. Hutson, T. W. Whaley, D. G. Ott, and C. T. Gregg, "Large-Scale Photosynthetic Production of Carbon-13 Labeled Sugars: The Tobacco Leaf System," *Biochem. Biophys. Res. Commun.*, **50**, 826-831 (1973); (b) V. H. Kollman, C. T. Gregg, J. L. Hanners, T. W. Whaley, and D. G. Ott, "Large-Scale Photosynthetic Production of Carbon-13 Labeled Sugars," in reference 10b, pp. 30-40; (c) V. H. Kollman, R. E. London, J. L. Hanners, and T. E. Walker, "Photosynthetic Preparation of ^{13}C-Labeled Sugars Using a Blue-Green Alga," in reference 10c, pp. 19-27.

[29] R. E. London, C. T. Gregg, and N. A. Matwiyoff, "Carbon-13 Nuclear Magnetic Resonance Study of the Rotational Mobility of Intra- and Extracellular Mouse Hemoglobin Labeled with ^{13}C Enriched Histidine," *Science*, **188**, 266-268 (1975).

[30] K. Yamada, S. Kinoshita, T. Tsunoda, and K. Aida, Eds., *The Microbial Production of Amino Acids*, Wiley, New York, 1972.

[31] J. H. Fletcher, O. C. Dermer, and R. B. Fox, Eds., *Nomenclature of Organic Compounds. Principles and Practice*, Advances in Chemistry Series, R. F. Gould, Ed., American Chemical Society, Washington, DC, 1973.

[32] Commission on Nomenclature of Organic Chemistry, "Nomenclature of Organic Chemistry. Section H. Isotopically Modified Compounds (First Edition)," *Pure and Appl. Chem.*, **51**, 351-380 (1979).

2

Acids, Anhydrides, Amides, Esters, and Nitriles

The carboxylic acids and derivatives comprise the largest group of labeled compounds. The isotopes are relatively readily introduced into the functional groups using the basic labeled starting materials—carbon-^{13}C monoxide, carbon-^{13}C dioxide, ammonia-^{15}N, and water-^{18}O—or nonfunctionally labeled compounds can be prepared from labeled organic precursors. In addition to the materials often being of ultimate use themselves, they are also important synthetic intermediates owing to the many possibilities for further reactions and transformations of the functional groups. The lower members of the series, in particular, provide the building blocks for construction of more complex molecules labeled at specific locations.

Carbon-13 can be directly introduced into the carboxyl group by way of the Grignard reaction with carbon-^{13}C dioxide or through carbonylation with carbon-^{13}C monoxide. The Grignard (or related organometallic) reactions are the more common for reasons that include wide applicability, familiarity, experience (especially carbon-^{14}C labeling), and utilization of standard laboratory apparatus. Carbonylation involves apparatus and gas handling techniques that are usually less familiar to most synthetic chemists, possibilities for isomerization, and fewer proven examples. Carbon-^{13}C monoxide is, however, the fundamental form for this isotope, and, as experience is gained and newer methods (such as polymer-bound rhodium catalysts, organopalladium substrates, and novel organometallic reactions) are applied, it can be expected to become a more popular starting material. Hydrolysis of nitriles, synthesized from cyanide-^{13}C, is a widely used and efficient method, although it is more indirect since the cyanide has to be first produced from carbon-^{13}C monoxide. Choice of a particular scheme (as for any synthesis) depends on a number of factors—compatability with other functional groups present, available starting materials and apparatus, experience, economics, and so on.

16

Nitrogen-15 cyanide has been prepared only rarely. Nitriles labeled with this isotope are usually obtained by dehydration of amides that are synthesized from ammonia-^{15}N and an appropriate acid derivative.

Oxygen-18 labeled acids, esters, and amides are usually prepared by hydrolysis of an appropriate compound (ester, orthoester, nitrile, imidate, trichloromethyl group) with water-[18]O. Labeling of specific sites, when a choice exists, can be accomplished through consideration of reaction mechanisms (which often were originally determined with the aid of oxygen-18). The possibilities for exchange reactions, which may either be desirable or undesirable, must also be recognized.

New synthetic methods that are continually being developed, such as the following examples, often appear attractive for isotopic syntheses. Palladium-catalyzed carbomethoxylation of olefins[1] that takes place with methanol, carbon monoxide, cupric chloride, and buffer (for example, dimethyl succinates from ethylenes) looks very promising for carbon-13 labeling schemes. Direct carbamoylation of allyl bromides with bis(N,N-diethylcarbamoyl)cuprate (from carbon monoxide, cuprous chloride, and lithium diethylamide) affords N,N-diethyl-3-butenamides; acyl halides give the 2-oxoamides.[2] Phosgene, a versatile but seldom used intermediate (probably because of lack of a readily applied preparative method) should be conveniently labeled in high yield through reaction of chlorine with carbon monoxide in the presence of tributylphosphine oxide.[3] More convenient and efficient variations of established methods are continually being developed that can readily replace older procedures, such as dehydration of amides to nitriles with trifluoroacetic anhydride[4] and an attractively simple preparation of esters from carboxylate salts and alkyl halides.[5] Amino acids have always been and will continue to be a most important class of labeled compounds, and many schemes have been devised for incorporation of the isotopes. Applications to labeling of recent developments,[6] particularly with respect to asymmetric synthesis, should provide desired isomers more efficiently and in higher yields. Utilization of supported catalysts[7] in carbonylation reactions should prove advantageous and, additionally, insulate the apparatus from highly corrosive reagents.

References

[1]J. K. Stille and R. Divakaruni, "Palladium(II)-Catalyzed Carboxylation Reactions of Olefins: Scope and Utility," *J. Org. Chem.*, 44, 3474-3482 (1979).

[2]T. Tsuda, M. Miwa, and T. Saegusa, "Lithium Bis(N,N-diethylcarbamoyl)cuprate. A Reagent for Direct Carbamoylation," *J. Org. Chem.*, 44, 3734-3736 (1979).

[3]M. Masaki, N. Kakeya, and S. Fijimura, "A Noval Catalytic Effect of Tertiary Phosphine Oxides and Dichlorides for the Reaction of Chlorine with Carbon Monoxide. A Preparative Method of a Phosgene Solution," *J. Org. Chem.*, 44, 3573-3574 (1979).

[4]F. Campagna, A. Carotti, and G. Casini, "A Convenient Synthesis of Nitriles from Primary Amides," *Tetrahedron Lett.*, 1977, 1813-1816.

[5]G. G. Moore, T. A. Foglia, and T. J. McGahan, "Preparation of Hindered Esters by the Alkylation of Carboxylate Salts with Simple Alkyl Halides," *J. Org. Chem.*, 44, 2425-2429 (1979).

[6]S. Yamada, T. Oguri, and T. Shioiki, "Asymmetric Synthesis of α-Amino-acid Derivatives by Alkylation of a Chiral Schiff Base," *J. Chem. Soc. Chem. Commun.*, 1976, 136-137; P. Vey

and J. P. Vevert, "α-Alkylation of α-Amino Acids," *Tetrahedron Lett.*, **1977**, 1455; L. Duhamel and J.-C. Plaquevent, "Deracemization by Enantioselective Protonation. A New Method for the Enantiomeric Enrichment of α-Amino Acids," *J. Am. Chem. Soc.*, **100**, 7415-7416 (1978).
[7]R. H. Grubbs, "Hybrid-Phase Catalysts," *CHEMTECH*, **1977**, 512-518.

FORMIC-^{13}C ACID

$$*CO + NaOH \xrightarrow[H_2O]{170°} H*COONa \xrightarrow[\text{2. } B_2O_3]{\text{1. } H_3PO_4} H*COOH$$

$$H*COONa \xrightarrow[170\text{-}180°]{(EtO)_3PO} H*COOEt$$

Procedure

Sodium formate-^{13}C[1,2] Into a 300-ml pressure vessel (Note 1) containing a magnetic stirring bar are placed sodium hydroxide (25%, 18 ml, 0.15 mol) and carbon-^{13}C monoxide (180 psi, 0.15 mol). The mixture is magnetically stirred and heated in an oil bath at 170° until the pressure decreases to a constant value (Note 2). After cooling, the vessel is opened, the reaction mixture is transferred to a round-bottom flask, and the water is evaporated (Note 3) to give the product as a white, crystalline solid (10.2 g, 99% yield) (Note 4).

Formic-^{13}C acid[1] To sodium formate-^{13}C (8.27 g, 0.12 mol) is added phosphoric acid (85%, 16 ml, 0.24 mol). Bulb-to-bulb vacuum distillation gives a mixture of the product and water (Note 5). Treatment with boric anhydride (20 g, 0.3 mol) and bulb-to-bulb vacuum distillation affords anhydrous formic-^{13}C acid (5.35 g, 95% yield) (Note 6).

Ethyl formate-^{13}C Sodium formate-^{13}C (3.80 g, 56.0 mmol) and triethyl phosphate (13 ml, 80 mmol) are placed in a 100-ml 3-neck flask fitted with an argon inlet, thermometer, air-cooled spiral condenser, and magnetic stirrer. The outlet of the condenser connects to two Dry Ice traps, followed by a Drierite tube and silicone oil bubbler. With argon flushing (approx. 5 ml/min), the flask is heated in an oil bath to 170-180° (Note 7) for several hours until reaction is complete (Note 8). The first cold trap contains all the product (3.80 g, 92% yield) (Note 9).

Notes

1 Either an autoclave or a stainless steel cylinder can be used. Smaller capacity, with consequent higher initial pressure and reaction rate, is somewhat advantageous, if the necessary higher initial carbon monoxide pressure is available.

2 The pressure approximately doubles on heating, and then decreases over 12-24 hr to about 40 psig (vapor pressure of water).

3 Removal of the last traces of water is aided by additions of isopropyl alcohol and reevaporation.

4 The product is satisfactory for further reactions, such as to the acid, esters, or to oxalic-^{13}C$_2$ acid.[2]

5 Although use of 85% (syrupy) phosphoric acid results in the distilled formic-^{13}C acid containing water, the method is clean and simple. Hydrogen chloride, with a flow system and appropriate cold traps, can also be used if attention is given to proper trap temperatures and pressures for distilling the final product. Sulfuric acid decomposes sodium formate to carbon monoxide, even at low temperature.

6 ^{1}H-NMR analysis shows the only impurity to be a trace of water, and the ^{13}C-concentration to be the same as the starting carbon-^{13}C monoxide.

7 Onset of the reaction occurs at about 170° as evidenced by bubbles in the reaction mixture and refluxing in the condenser. The (air-cooled) condenser serves primarily as a spray trap. Although its temperature remains below the boiling point of ethyl formate, vapor pressure of the ester and the argon flow are sufficient to carry the product (quantitatively) into the cold trap. Use of hexamethylphosphoramide as a solvent lowers the reaction temperature to about 150°; however, this is of no great advantage, and the product has an odor of methylamine (although pure by ^{1}H-NMR). Diethyl sulfate produces the ester, but in lower yield with concomitant generation of carbon monoxide and diethyl ether (the ether is also produced by heating diethyl sulfate with sodium carbonate).

8 Heating is continued for awhile after condensate is no longer present in the condenser.

9 ^{1}H-NMR analysis indicates the product to be pure. Owing to the volatility of the ester, care should be taken to avoid loss in transferring from the trap.

Other Preparations

Sodium formate-^{13}C has been prepared by ethanolic sodium hydroxide hydrolysis of isopropyl formate-^{13}C.[3]

References

[1]D. G. Ott, in "Annual Report of the Biomedical and Environmental Research Program of the LASL Health Division," compiled by E. C. Anderson and E. M. Sullivan, Los Alamos Scientific Laboratory Report LA-5883-PR, February 1975, p. 147.

[2]B. D. Andresen, "Synthesis of Sodium Formate-^{13}C and Oxalic Acid-^{13}C$_2$," *J. Org. Chem.*, **42**, 2790 (1977).

[3]R. E. Royer, G. H. Daub, and D. L. Vander Jagt, "Synthesis of Carbon-13 Labelled 6-Substituted Benzo[a]-pyrenes," *J. Labelled Compds.*, **12**, 377-380 (1976).

FORMAMIDE-^{13}C-^{15}N

$$*CO + Me_2CHOH \xrightarrow{Me_2CHONa} H*COOCHMe_2$$

$$\xrightarrow[Me_2CHOH]{{}^\bullet NH_3} H*CO^\bullet NH_2$$

Procedure

Isopropyl formate-^{13}C[1,2] Sodium (1.0 g, 43 mmol) is dissolved in isopropyl alcohol (380 ml), and the solution is transferred to a 1-liter stirred autoclave (Note 1). Carbon-^{13}C monoxide (806 psi, 1.4 mol) is added, and the mixture is stirred at room temperature until the pressure falls to an equilibrium value (approx. 56 psig after 48 hr). After venting (Note 2) the solution is treated with anhydrous Dowex-50(H$^+$) (150 ml) for 1 hr (Note 3), filtered, and fractionally distilled (52-53° at 580 mm Hg) through a spinning-band column to give the azeotrope (114.5 g) containing 90 wt.% of isopropyl formate-^{13}C (103.1 g, 89% yield) (Note 4).

Formamide-^{13}C[1] The above-mentioned solution of isopropyl formate-^{13}C, isopropyl alcohol, and sodium isopropoxide is transferred to a 1-liter 3-neck flask equipped with a magnetic stirrer, Dry Ice condenser with drying tube, and an inverted 300-cm^3 steel cylinder containing liquid ammonia (approx. 60 g). The solution is cooled to −70°, the ammonia is added, and the mixture is stirred for 3 hr after allowing the temperature to rise slowly to −30°. The excess ammonia is evaporated by replacing the Dry Ice condenser with a water-cooled condenser and stirring the mixture overnight at room temperature. The solution is treated with anhydrous Dowex-50(H$^+$) (30 g) to remove sodium ions, filtered, and distilled at atmospheric pressure to remove most of the alcohol. Fractional distillation of the remainder gives formamide-^{13}C (55.0 g, 92% yield from carbon-^{13}C monoxide), bp 138-140° at 60 mm Hg.

Formamide-^{13}C-^{15}N[3] This isotope isomer is prepared in the above manner, except that the ammonia-^{15}N (in 10% excess, Note 5) is distilled into the crude isopropyl formate-^{13}C mixture. The reaction vessel is closed, and the contents are stirred for 3 days at room temperature. The product is isolated as above (88% yield from ammonia-^{15}N).

Formamide-^{15}N Ethyl formate (Note 6) (7.7 g, 0.12 mol) and ethanol (25 ml) are placed in a 75-cm^3 stainless steel gas sampling cylinder containing a magnetic stirring bar. The cylinder is fitted with a valve that connects to the vacuum and to a storage vessel containing ammonia-^{15}N (1.80 g, 10.0 mmol), cooled in liquid nitrogen, and evacuated. The ammonia is transferred to the reaction mixture, and the closed cylinder is magnetically stirred at room temperature for 3 days (Note 7). The valve is removed from the cylinder, and the solution is transferred, with ethanol rinses, to a distilling flask. After distillation of most of the ethanol, the product is transferred to a smaller flask for vacuum distillation, which affords the product (4.20 g, 91% yield) (Note 8).

Notes

1 A stainless steel gas cylinder of appropriate size for the desired scale, containing a magnetic stirring bar, can serve as the reaction vessel.

2 The unreacted carbon-^{13}C monoxide can be recovered by venting through a Dry Ice trap into a liquid nitrogen-cooled storage cylinder. The amount unreacted

(5-20%) appears to be related to the presence of water in the reaction mixture.

3 Anhydrous resin is prepared by repeated evaporations at reduced pressure of a slurry of the resin in benzene or toluene.

4 The crude solution can be used directly for conversion to formamide; prior to distillation, however, the base must be removed or reversal of the reaction with decomposition to carbon monoxide will occur.

5 A stoichiometric amount of ammonia should be satisfactory or, perhaps, better than an excess.

6 The methyl or isopropyl esters may also be used.

7 Progress of the reaction can be followed by fitting a pressure gage to the reaction vessel.

8 The loss is almost entirely from holdup in the distillation apparatus; additional product can be recovered (in solution) by rinsing.

Other Preparations

Methyl formate-^{13}C has been prepared by the same procedure starting with methanol in place of isopropyl alcohol; care must be taken to avoid loss because of its greater volatility.

References

[1] T. W. Whaley and D. G. Ott, "Syntheses with Stable Isotopes: Sodium Cyanide-^{13}C," *J. Labelled Compds.*, **11**, 307-312 (1975).

[2] S. D. Larsen, P. J. Vergamini, and T. W. Whaley, "Synthesis of Cyclopentadienyl-x-^{13}C Thallium," *J. Labelled Compds.*, **11**, 325-332 (1975).

[3] D. G. Ott, in "Annual Report of the Biomedical and Environmental Research Program of the LASL Health Division," compiled by E. C. Anderson and E. M. Sullivan, Los Alamos Scientific Laboratory Report LA-5883-PR, February 1975, p. 129.

DIMETHYLFORMAMIDE-^{18}O

$$HCONMe_2 + PhCOCl \longrightarrow PhCOOCH=NMe_2$$

$$\xrightarrow[0\text{-}5^\circ]{H_2{}^*O} HC^*ONMe_2 + PhCOOH + HCl$$

Procedure[1]

A mixture of dimethylformamide (7.3 g, 0.10 mol) (dried over potassium hydroxide and distilled) and benzoyl chloride (14.0 g, 0.10 mol) in an Erlenmeyer flask equipped with a magnetic stirrer and a drying tube is cooled to 0-5°, and water-^{18}O (1.85 g, 0.10 mol) is introduced slowly while stirring. After about 5 min the mixture solidifies owing to precipitation of benzoic acid (Note 1). Sodium bicarbonate (8.4 g, 0.10 mol) is added, and after evolution of the carbon dioxide the mixture is extracted with acetone (3 × 15 ml). The acetone solution is dried over potassium carbonate (5 g), filtered, and evaporated at reduced pressure. The residual liquid is distilled at atmospheric pressure to give the product (6.0 g, 81% yield), bp 153° (Note 2).

Notes

1 Analysis of the benzoic acid, isolated by filtration or extraction, shows the absence of oxygen-18.
2 Mass spectrometric analysis shows 90% incorporation of the oxygen-18; that is, the product from water-^{18}O of 20 mol % ^{18}O contains 18 mol % ^{18}O. Equimolar quantities of water-^{18}O and dimethylformamide when heated for 24 hr at 60° with hydrochloric acid show no isotopic exchange.

References

[1]R. R. Koganty and G. A. Digenis, "A Convenient Preparation of ^{18}O-Dimethylformamide," J. Labelled Compds., 10, 419-422 (1973).

N-METHYL-N-PHENYLFORMAMIDE-^{13}C (N-METHYLFORMANILIDE-1-^{13}C)

$$H*COOCHMe_2 \xrightarrow[\text{EtOH}]{\text{NaOH}} H*COONa \xrightarrow[150°]{\text{PhNMe·HCl}} PhNMe*CHO$$

Procedure[1]

A solution containing isopropyl formate-^{13}C (38.3 g, 0.35 mol) (Note 1), 100 ml of 40% aqueous ethanol, and sodium hydroxide (0.35 mol) is refluxed for 2 hr. To the sodium formate-^{13}C remaining after evaporation of the solvents is added N-methylaniline hydrochloride (65 g, 0.45 mol), and the mixture is heated at 150° for 1 hr. The water formed is removed by distillation, hydrochloric acid (5%, 50 ml) is added, and the solution is extracted with ether (4 × 50 ml). The ether solution is dried over magnesium sulfate and evaporated, and the product (36.4 g, 76% yield) is distilled, bp 114-121° at 8 mm Hg.

Notes

1 The ester used also contained a small amount of isopropyl alcohol.

References

[1]R. E. Royer, G. H. Daub, and D. L. Vander Jagt, "Synthesis of Carbon-13 Labelled 6-Substituted Benzo[a]pyrenes," J. Labelled Compds., 12, 377-380 (1976).

POTASSIUM OR SODIUM CYANIDE-^{13}C

Method I

$$H*CONH_2 \xrightarrow[\text{THF}]{Ph_3P, CCl_4} H*CN \xrightarrow{NaOH} Na*CN$$

Procedure (Note 1)

A 1-liter 3-neck flask equipped with a magnetic stirrer, nitrogen inlet, addition funnel, and condenser connected to a trap containing sodium hydroxide (5 M, 50 ml) (Note 2) is charged with triphenyl phosphine (131 g, 0.50 mol) and carbon tetrachloride (375 ml) and heated to 40°. A solution of formamide-^{13}C (11.5 g, 0.25 mol) in tetrahydrofuran (375 ml) is added from the funnel over a 2-hr period with nitrogen flushing. After approx. 60 hr, during which vigorous stirring, nitrogen flow, and reaction temperature of 40-45° are maintained, the solution from the trap (Note 3) that contains the sodium cyanide-^{13}C (0.20 mo. 80% yield) is transferred to a dropping funnel and added slowly, with magnetic stirring, to sulfuric acid (30%, 250 ml) (Note 4). The evolved hydrogen cyanide-^{13}C is collected, with nitrogen flushing, in a Dry Ice-cooled 500-ml flask, dissolved in cold water, analyzed, and neutralized with an equivalent of sodium hydroxide solution. Lyophilization (Note 5) gives solid sodium cyanide-^{13}C (Note 6).

Method II

$$*CH_4 \xrightarrow[\text{Pt, } 1000°]{NH_3} NH_4*CN \xrightarrow{NaOH} Na*CN$$

Procedure (Note 7)

Methane-^{13}C (3.2 g, 0.20 mol) at 0.07-0.09 liter/min and excess ammonia at 0.45 liter/min are passed through a quartz reaction tube (1.2 × 90 cm) containing 4 sheets of rolled and crumpled platinum foil (7.8 g) distributed along 80 cm within the tube and heated to 1000° by an electric tube furnace (Note 8). The exit gases are passed through an air-cooled spiral condenser into an ice/alcohol-cooled trap at −20° and then through sodium hydroxide solution (5 M, 30 ml). The contents of the sodium hydroxide trap (containing 0.01 mol of cyanide) are combined with the ammonium cyanide (7.5 g, 0.17 mol) in the cold trap, additional sodium hydroxide (0.03 mol) is added, and the solution is lyophilized (Note 5) to give the product as a white powder (9.4 g, 90% yield) (Note 6).

Method III

$$*C \xrightarrow[\text{2. KOH}]{1. \ NH_3, 1100°} K*CN$$

Procedure (Note 9)

Carbon-^{13}C (1.3 g, 0.10 mol) (Note 10) is placed in a fused-silica boat in a fused-silica combustion tube that is heated to 1100°. A mixture of ammonia at approx. 70 ml/min and argon at 20 ml/min is passed over the carbon and then through potassium hydroxide solution (0.4 M, 247 ml). The reaction is stopped after 87 mmol of cyanide is absorbed (Note 11), and the solution is evaporated to a concentration of 0.83 M (91% yield).

Notes

1   This method[1] is the best of several formamide dehydration reactions studied (phosphorus pentoxide-toluene, phosphorus pentoxide-pyridine, toluenesulfonyl chloride-pyridine, acetic anhydride, trifluoroacetic anhydride-pyridine, phosphorus oxychloride, and vapor phase over alumina). It has been applied on scales of a few millimoles to a mole with little variation in yield.

2   Choice between the sodium or potassium derivative is often only historical and immaterial; for certain applications considerations should be given to differences in reactivity.[2]

3   Aliquots from the trap are periodically analyzed by titration with standard silver nitrate (potassium iodide indicator).

4   Purification of the crude product by regeneration of hydrogen cyanide is necessary to eliminate small amounts of colored impurities and excess sodium hydroxide (see Note 5).

5   Evaporation of aqueous cyanide solutions should be carried out at low temperature and pressure to avoid loss of hydrogen cyanide. It is suggested[3] that the tedious water-evaporation step could be avoided by trapping the product in cold methanol, adding an equivalent of sodium or potassium hydroxide, and isolating by vacuum evaporation of the alcohol. Also, use of sulfuric acid traps preceding the product trap in the dehydration reaction should eliminate the need for regeneration.

6   Purity is approx. 96% by argentimetric analysis; the remainder is presumed to be sodium hydroxide and water (similar to commercial, nonisotopic, material).

7   This procedure[3] was developed primarily for large-scale preparations (several moles) for which it is more convenient (once the apparatus has been assembled) than Method I. A yield of 93% is reported from 1.15 mol of methane-^{13}C using a 2.5 × 90 cm reaction tube packed with 276 g of platinum as perforated sheet and gauze.

8   The reaction tube is brought to temperature while flushing with helium. Passage of the reactants requires about 2 hr. Proper flow rate should be determined in trial runs utilizing infrared analysis for absence of methane in the exit gas.

9   This method[4] does not require the platinum catalyst nor control of flow rates as does Method II and is quite convenient if elemental carbon-^{13}C is available.

10   The carbon-^{13}C should contain less than 500 ppm iron.

11   The reaction, which requires 1-2 hr, is followed by titration of aliquots from the trap. At termination 1.16 g (89.6 mmol) of the carbon-^{13}C had reacted.

Other Preparations

Many of the schemes developed for cyanide-^{14}C,[5] are often applicable on a scale of only a few millimoles (owing to their somewhat hazardous nature). Recently, however, 50-mmol quantities of potassium cyanide-^{13}C have been prepared from barium carbonate-^{13}C, potassium, and ammonium chloride.[6] By scaling up a procedure for cyanide-^{14}C,[5] silver cyanide-^{13}C has been isolated in 73% overall yield from carbon-^{13}C dioxide following hydrogenolysis of triphenylacetonitrile-^{13}C with sodium and ethanol.[7] Sodium cyanide-^{13}C-^{15}N and cyanide-^{15}N have been obtained by Method I from formamide-^{13}C-^{15}N and cyanide-^{15}N and formamide-^{15}N, respectively.[8] Although nitrogen-15 labeling by Methods II and III would require recovery and recirculation of ammonia-^{15}N, this might nevertheless be practical for large-scale preparations.

References

[1] T. W. Whaley and D. G. Ott, "Syntheses with Stable Isotopes: Sodium Cyanide-^{13}C," *J. Labelled Compds.*, 11, 307-312 (1975).

[2] J. E. Shaw, D. Y. Hsia, G. S. Parries, and T. K. Sawyer, "Reaction of Alkali Metal Cyanides with Alkyl Halides in HMPA or HMPA Containing Crown Ether," *J. Org. Chem.*, 43, 1017 (1978).

[3] D. G. Ott, V. N. Kerr, T. G. Sanchez, and T. W. Whaley, "Syntheses with Stable Isotopes: Sodium Cyanide-^{13}C, Methane-^{13}C, Methane-^{13}C-d$_4$, and Methane-d$_4$," *J. Labelled Compds.*, 17, 255-262 (1980).

[4] W. M. Rutherford and J. C. Liner, "Preparation of Elemental ^{13}C and K^{13}CN from ^{13}CO$_2$," *Int. J. App. Radiat. and Isot.*, 21, 71-73 (1970).

[5] A. Murray and D. L. Williams, *Organic Syntheses with Isotopes*, Interscience, New York, 1958, pp. 563-567.

[6] B. Stuetzel, W. Ritter, and K-F. Elgert, "^{13}C-Enriched Monomers for NMR-Spectroscopy of Polymers," *Angew. Makromol. Chem.*, 50, 21-41 (1976).

[7] A. Murray III, in "Annual Report of the Biological and Medical Research Group (H-4)," C. R. Richmond and G. L. Voelz, Eds., Los Alamos Scientific Laboratory Report LA-4923-PR, April 1972, p. 123.

[8] D. G. Ott, in "Annual Report of the Biomedical and Environmental Research Program of the LASL Health Division," compiled by E. C. Anderson and E. M. Sullivan, Los Alamos Scientific Laboratory Report LA-5883-PR, February 1975, p. 129.

UREA-^{13}C, UREA-^{12}C, OR UREA-^{13}C-^{15}N$_2$

$$*CO + 2\,{}^\bullet NH_3 + S \xrightarrow{\text{MeOH}} H_2{}^\bullet N*CO{}^\bullet NH_2 + H_2S$$

Procedure[1]

Urea-^{13}C or urea-^{12}C A stainless steel, 3-liter gas cylinder with a dial thermometer in one end is charged with hydrogen sulfide-saturated methanol (150 g) and sulfur (32 g, 1.0 mol), closed with a T-fitting bearing a pressure gauge and valve, and cooled in Dry Ice. Ammonia (45 g, 2.6 mol) is cryogenically transferred to the reaction mixture followed by carbon-[13]C (or -^{12}C) monoxide (20.5 g, 0.71 mol) (Note 1). The reaction cylinder is wrapped with heating tape and insulation, fastened to a mechanical shaker, and heated to 100°. After 6 hr (Note 2) the cylinder is cooled, vented, drained, and rinsed with warm methanol. Evaporation of the dark-red solution at reduced pressure gives a green-gray solid (55 g) that is extracted with warm water (250 ml). The extract is evaporated at reduced pressure to give the crude product (41 g), which is recrystallized from ethanol (using activated carbon) as colorless needles, mp 134-135° (37.5 g, 87% yield).

Urea-[13]C-[15]N$_2$ The same procedure, employing a 1-liter reaction cylinder, sulfur (10.6 g, 0.33 mol), hydrogen sulfide-saturated methanol (50 ml), a methanol solution of ammonia-^{15}N, and carbon-^{13}C monoxide (9.7 g, 0.34 mol), gives urea-[13]C-[15]N$_2$ (5.0 g, 34% yield, Note 4).

Notes

1 The gases are transferred from cylinders containing the desired quantities of each gas obtained by previous cryogenic transfers.

2 Maximum pressure of 200 psig is reached as the temperature reaches 100° and declines thereafter. Infrared analysis of a gas sample shows no carbon monoxide; after 3 hr 18% remains unreacted.

3 The ammonia-^{15}N, generated by refluxing a slurry of ammonium-^{15}N chloride (24.7 g, 0.45 mol) in methanol (200 ml) with a solution of sodium (16 g, 0.7 mol) in methanol (160 ml), is flushed with helium into liquid nitrogen-cooled traps to give a methanol-ammonia solution (20 g). The yield, determined in trial runs, is 96-100%. The entire solution is cryogenically transferred to the reaction cylinder.

4 The low yield is attributed to a mechanical loss; trial runs gave crude urea in 83% yield.

References

[1]T. W. Whaley and D. G. Ott, "Syntheses with Stable Isotopes: Urea-^{13}C, Urea-^{12}C, and Urea-[13]C-[15]N$_2$," *J. Labelled Compds.*, **11**, 167-170 (1975).

DIETHYL CARBONATE-[13]C

$$Ba*CO_3 \xrightarrow[H_2O]{AgNO_3} Ag_2*CO_3 \xrightarrow[DMF]{EtI, Et_3N} (EtO)_2*CO$$

Procedure[1]

Silver carbonate-[13]C Finely powdered barium carbonate-[13]C (1.97 g, 10.0 mmol) is added to a solution of silver nitrate (20 g, 0.12 mol) in water (50 ml), and the mixture is magnetically stirred overnight (Note 1). The product is filtered, washed with water (3 X 10 ml), acetone, and ether, and air dried (95-100% yield).

Diethyl carbonate-[13]C (Notes 1 and 2)   A mixture of dimethylformamide (10 ml), ethyl iodide (5.0 ml, 62 mmol), and triethylamine (2.0 ml, 14 mmol) is allowed to stand about 20 min until tetraethylammonium iodide begins to precipitate. Freshly prepared silver carbonate-[13]C (10 mmol) is added, the mixture is stirred for 6-12 hr, additional triethylamine (6.8 ml, 49 mmol) is added, and stirring is continued for 6-12 hr. The mixture is treated with cyclohexane (20 ml), stirred 2 hr, and filtered. The precipitate is washed with dimethylformamide-cyclohexane (7 ml:20 ml), and the filtrate is treated with phosphate buffer (0.1 M, pH 7, 10 ml) and saturated sodium chloride (10 ml) and again filtered. The upper phase, dried over sodium sulfate, contains cyclohexane, traces of dimethylformamide and triethylamine, and the product (60-85% yield) (Note 3).

Notes

1   The flask is wrapped with aluminum foil to exclude light. All manipulations are conducted in the dark or subdued light.
2   Effects of solvent, temperature, times, metal ions, and triethylamine were studied.
3   The cyclohexane solution can often be used without further purification. A second cyclohexane extraction of the aqueous phase affords little additional product, but increases the level of contaminants.

References

[1]W. R. Porter, L. A. Spitznagle, and W. F. Trager, "A High Yield Synthesis of [13]C- and ^{14}C-Labeled Diethyl Carbonate," *J. Labelled Compds.*, **12**, 577-582 (1976).

[(CYANO-^{13}C)ACETYL] UREA

$$K*CN + ClCH_2COONa \xrightarrow[\text{2. HCl}]{\text{1. H}_2\text{O}} HOOCCH_2*CN$$

$$\xrightarrow[\text{Ac}_2\text{O, 100}^\circ]{\text{H}_2\text{NCONH}_2} H_2NCONHCOCH_2*CN$$

Procedure[1]

Chloroacetic acid (3.46 g, 36.6 mmol) in the minimum amount of water (5.7 ml) is neutralized with sodium carbonate (1.95 g, 18.4 mmol) and mixed with a solution of potassium cyanide-[13]C (1.97 g, 30.3 mmol) dissolved in water at 60°. The temperature of the mixture is kept at 60° by immersion in cool water as necessary. After about 45 min the temperature ceases to rise, and the mixture is allowed to stand at room temperature for 12 hr. After acidification with hydrochloric acid (conc., 3 ml) the solution is evaporated at reduced pressure, and the syrupy residue is treated with alcohol (15 ml) and filtered to remove the salt. The solution is evaporated at reduced pressure, taken up in alcohol, and evaporated several times (Note 1). The dry residue of crude (cyano-^{13}C)acetic acid is treated with a suspension of urea (1.86 g, 31.0 mmol) in acetic anhydride (5 ml), and the mixture is heated at 100° for 30 min. Water (20 ml) is added and the solution is cooled to give the product, mp 211-212° (2.92 g, 76% yield) (Note 2).

Notes

1 The repeated evaporations, with temperature kept below 50°, assure complete removal of water.

2 The melting point, mixed melting point, and infrared spectrum are identical to those of an authentic sample.

References

[1]J. W. Triplett, S. W. Mack, S. L. Smith, and G. A. Digenis, "Synthesis of Carbon-13 Labelled Uracil, 6,7-Dimethyllumazine, and Lumichrome, via a Common Intermediate: Cyanoacetylurea," *J. Labelled Compds.*, 14, 35-41 (1978).

^{13}C-ACETIC ACID

Method I

$$*CO_2 + {}^\bullet CH_3MgI \xrightarrow[\text{2. H}^+, \text{H}_2\text{O}]{\text{1. Et}_2\text{O}} {}^\bullet CH_3 *COOH$$

Procedure[1]

Acetic-1-^{13}C acid Into a multinecked reaction flask equipped with an efficient magnetically driven stirrer, thermometer, gas inlets, and manometer (Note 1) is introduced through a glass-wool filter a solution of methylmagnesium iodide, prepared from magnesium (12 g, 0.50 mol) and methyl iodide (71 g, 0.50 mol) in ether (750 ml). The Grignard reagent is flushed with helium, diluted with ether (800 ml), frozen with liquid nitrogen, and degassed by alternate thawing and freezing under vacuum. The frozen solution is evacuated to 0.5 mm Hg, isolated from the pump, and brought to −25° (Note 2). Carbon-^{13}C dioxide is admitted to the reaction flask following generation from barium carbonate-^{13}C (60 g, 0.30 mol) by addition of sulfuric acid (conc., 500 ml) at such a rate that the pressure in the reaction flask remains below 500 mm Hg. At the end of the reaction (Note 2) helium is admitted, and the mixture is hydrolyzed at −25° by addition of sulfuric acid (4 N, 100 ml). Silver sulfate (130 g) is added, and the entire mixture is transferred to a continuous extraction apparatus and extracted with ether for 2 days. The ether is distilled to give the crude product (20 g, 66% yield) (Note 3).

Acetic-2-^{13}C acid By the same procedure, but with a simplified apparatus, this isotope isomer is prepared from methyl-^{13}C iodide.[1]

Method II

$$*CO + \bullet CH_3OH \xrightarrow[\text{H}_2\text{O, 175}°]{\text{RhCl}_3, \text{HI}} \bullet CH_3 *COOH$$

Procedure[2]

Acetic-^{13}C$_2$ acid Methanol-^{13}C (143 g, 4.36 mol), water (120 g) (Note 4), rhodium trichloride (1.0 g, 5 mmol) (Note 5), and hydriodic acid (57%, 25 ml) are placed in a 1-gal, magnetically stirred autoclave of Hastelloy C alloy (Note 6). Carbon-^{13}C monoxide (415 psig, 4.47 mol) is added from a 300-cm^3 cylinder previously filled with the desired quantity (Note 7). The autoclave is electrically heated to 175° over a period of about 2 hr and kept at approximately that temperature until there is no further pressure drop (Note 8). After cooling to room temperature the residual pressure is vented through a Dry Ice trap, and the autoclave and trap contents are transferred to a distillation flask with several rinses of water. Following total distillation at reduced pressure (approx. 400 mm Hg) the solution is treated with phosphorous acid (30%, 5 ml) and silver sulfate (5 g) and redistilled at atmospheric pressure (Note 9). The forerun consists of methyl-^{13}C acetate-^{13}C$_2$ (8 g, 2% yield), acetic-^{13}C$_2$ acid (12 g, 4.5% yield), and water (25 g) (Note 10). The product fraction contains acetic-^{13}C$_2$ acid (243 g, 90% yield) and water (135 g). Anhydrous acid is obtained by treating the mixture with magnesium sulfate (1.25 g/g water) in small portions, filtering and washing the hydrated solid with dichloromethane, and distilling the solution (Note 11).

Acetic-^{12}C$_2$ acid This isomer is prepared by the same procedure from methanol-^{12}C and carbon-^{12}C monoxide (Note 12).

Acetic-1-^{13}C acid This isomer is prepared by the same procedure from methanol and carbon-^{13}C monoxide (Note 7).

Acetic-2-^{13}C acid This isomer is prepared by the same procedure from methanol-^{13}C, except that excess carbon monoxide is used (Note 7).

Notes

1 A drawing of the apparatus is given by Stuetzel, Ritter, and Elgert.[1]
2 The pressure is the vapor pressure of ether.
3 The product is suitable to use directly for various purposes, such as esterification and salt formation.
4 The amount of water is not critical.
5 Rhodium trichloride trihydrate may also be used. Other rhodium compounds have not been studied for this purpose, but should be applicable.

6 Stainless steel vessels are not advisable owing to the corrosive nature of rhodium carbonyl compounds. The reaction is readily carried out on smaller scales by using smaller autoclaves.

7 For the double-labeled or carboxyl-labeled isomers a compromise is necessary with respect to amount of carbon-^{13}C monoxide used, methyl acetate produced, and overall yield. With methyl labeling carbon monoxide is utilized in excess, and methyl acetate is not a product.

8 As the temperature reaches 175° the pressure is maximum at about 800 psig, and the reaction (exothermic) proceeds over the next several hours with decline of the pressure to a steady value of approx. 150 psig. On cooling to room temperature only a slight pressure of carbon monoxide remains.

9 Phosphorous acid reduces the iodine present, and silver sulfate immobilizes the iodide.

10 Analyses are by gas chromatography and ^{1}H-NMR.

11 This procedure removes water from the product more efficiently and conveniently than other methods, such as azeotropic distillation with chloroform and carbon tetrachloride, treatment with molecular sieves, or regeneration of the acid following conversion to a salt. It is likely that treatment with boric anhydride (see Formic-[13]C Acid) would also be effective.

12 Care must be taken to use scrupulously clean apparatus to avoid contaminating the product with carbon-13.

Other Preparations

Acetic-1-^{13}C acid has been liberated from sodium acetate-1-^{13}C, prepared by a procedure similar to Method I, by treating the salt with polyphosphoric acid and heating to 170-180°.[3] Phosphorus pentoxide and water in molar ratio of 1:3.03 with anhydrous sodium acetate-1-^{13}C gives the free acid in quantitative yield, but it contains approx. 11 wt. % water.[4] Treatment of the potassium salt with phosphorus pentoxide-saturated 84% phosphoric acid provides the anhydrous acid in quantitative yield.[5]

By a procedure similar to Method I octanoic-1-^{13}C acid has been prepared[6] starting from heptylmagnesium bromide, although isolation of the higher molecular weight product is simpler. The crude product from the acidified Grignard reaction is converted to its sodium salt, washed with chloroform, treated with hydrochloric acid, and extracted into chloroform. The solvent is evaporated to give material showing no impurities by IR or NMR. The acid is esterified by treatment with excess iodomethane and potassium carbonate in hexamethylphosphoramide to give methyl octanoate-1-^{13}C, which is isolated by extracting into carbon tetrachloride, washing with water, and distilling at reduced pressure. Transesterification with triacetin, catalyzed by sodium methoxide, gives the ^{13}C-trioctanoin [1,2,3-propanetriyl tri(octanoate-1-^{13}C)].

References

[1]B. Stuetzel, W. Ritter, and K-F. Elgert, "^{13}C-Enriched Monomers for NMR-Spectroscopy of Polymers," *Angew. Makromol. Chem.*, **50**, 21-41 (1976).

[2] D. G. Ott and V. N. Kerr, "Syntheses with Stable Isotopes: Acetic-1-^{13}C, -2-^{13}C, -^{13}C$_2$, and -^{12}C$_2$ Acids," *J. Labelled Compds.*, **12**, 119-125 (1976).

[3] D. L. Fitzell, D. P. H. Hsieh, C. A. Reece, and J. N. Seiber, "Preparation of Acetic-1-^{13}C and Malonic-1-^{13}C Acids for Biosynthetic Studies," *J. Labelled Compds.*, **11**, 135-139 (1975).

[4] J. L. Roberts and C. D. Poulter, "2′,3′,5′-Tri-O-benzoyl[4-^{13}C]uridine. An Efficient, Regiospecific Synthesis of the Pyrimidine Ring," *J. Org. Chem.*, **43**, 1547-1550 (1978).

[5] B. Bak and J. J. Led, "Preparation of [1-^{13}C], [3-^{13}C], [1-D] and [3-D] Enriched Cyclobutenes," *J. Labelled Compds.*, **4**, 22-27 (1968).

[6] J. B. Watkins, D. A. Schoeller, P. D. Klein, D. G. Ott, A. D. Newcomer, and A. F. Hofmann, "^{13}C-Trioctanoin: A Nonradioactive Breath Test to Detect Fat Malabsorption," *J. Lab. Clin. Med.*, **90**, 422-430 (1977).

ETHYL ACETATE-1-^{13}C OR ETHYL ACETATE-2-^{13}C

Method I

$$*CH_3{}^\bullet COOH \xrightarrow{\text{NaOH}} *CH_3{}^\bullet COONa \xrightarrow[170\text{-}210^\circ]{(EtO)_3PO} *CH_3{}^\bullet COOEt$$

Procedure[1]

A solution of acetic-1-^{13}C or acetic-2-^{13}C acid (2.00 g, 33.0 mmol) in water (15 ml) is neutralized (phenolphthalein end point) with sodium hydroxide, the solvent is evaporated at reduced pressure, and the pinkish crystals are dried at 140° and 0.1 mm Hg. The sodium acetate-1-^{13}C or acetate-2-^{13}C (2.71 g, quant. yield), mp 328-328.5°, is powdered and heated under reflux at 170-210° with triethyl phosphate (10 ml) and a piece of glass wool for 3 hr. The product is distilled at 0.1 mm Hg into a liquid nitrogen-cooled receiver (2.89 g, 100% yield) (Note 1).

Method II

$$*CH_3{}^\bullet COOH \xrightarrow[H_2SO_4]{\text{EtOH}} *CH_3{}^\bullet COOEt$$

Procedure[2]

Ethanol (5 ml) and sulfuric acid (5 ml) in a 250-ml 3-neck flask are stirred and heated at 140° while a solution of crude ^{13}C$_1$-acetic acid (20 g) (Note 2) in ethanol (20 ml) is added gradually from a dropping funnel. The ester distills during the course of the reaction, and the crude product is washed with sodium carbonate solution and then with calcium chloride (50%) solution (Note 3). After drying over sodium sulfate and calcium chloride the product is distilled, bp 75-79° (12 g) (Note 2).

Notes

1   ^{1}H-NMR data are given for the two isotope isomers.[1]

2   The quantity of ^{13}C-acetic acid in the crude starting material is not given. The overall yield of ethyl acetate-1-^{13}C is 46% from barium carbonate-^{13}C.

3   The washings remove acetic acid and ethanol that distill with the product.

References

[1]G. A. Braden and U. Hollstein, "Synthesis of 1-Phenyl-2-phenyl-1-^{13}C-ethene-1-^{13}C (trans-Stilbene) and Derivatives," *J. Labelled Compds.*, **12**, 507-516 (1976).

[2]B. Stuetzel, W. Ritter, and K-F. Elgert, "^{13}C-Enriched Monomers for NMR-Spectroscopy of Polymers," *Angew. Makromol. Chem.*, **50**, 21-41 (1976).

p-PHENYLPHENACYL ACETATE-^{18}O$_2$

$$H_2*O + CH_3C(OEt)_3 \xrightarrow{\text{TsOH}} CH_3C*OOEt \xrightarrow[\text{NaOMe}]{H_2*O}$$

$$CH_3C*O*ONa \xrightarrow[C_6H_6, CH_3CN, \text{cat.}]{p\text{-PhC}_6H_4COCH_2Br} CH_3C*O*OCH_2COC_6H_4Ph\text{-}p$$

Procedure[1]

Sodium acetate-^{18}O$_2$ Water-^{18}O (566.6 mg, 28.2 mmol), triethyl orthoacetate (40 ml) (Note 1), and p-toluenesulfonic acid monohydrate (2 mg) are stirred until homogeneous (approx. 1 min). To this ethyl acetate-CO-^{18}O solution is added an equivalent amount of sodium methoxide (2.63 N) in methanol (Note 2), followed by a second equivalent of water-^{18}O (577.3 mg, 29.1 mmol) in 1-2 ml of anhydrous tetrahydrofuran. The mixture, which solidifies to a fluffy, white crystalline mass within a few minutes, is heated on a water bath at 70° for 20 hr and then evaporated under reduced pressure at 70° (Note 3) to give the crystalline product (2.39 g, 91% purity, 89% yield).

p-Phenylphenacyl acetate-^{18}O$_2$ Sodium acetate-^{18}O$_2$ (10.2 mg, 0.125 mmol) is reacted with p-phenylphenacyl bromide (69 mg, 0.25 mmol) and 18-crown-6-acetonitrile (4 mg) in acetonitrile (1 ml) and benzene (1 ml) at reflux for 2 hr. The mixture is evaporated to an oil with a stream of nitrogen, dissolved in chloroform (2 ml) and water (2 ml), and extracted twice again with chloroform (3-ml portions). The combined organic extracts are washed with water and brine, dried over sodium sulfate, and evaporated to give a crystalline residue (92 mg), which is purified by preparative thin-layer chromatography on silica gel (benzene) to yield the product (60 mg, 18%, 91% purity) (Note 4).

Notes

1   The orthoacetate is dried by stirring over potassium carbonate for 20 hr and then distilled. Glassware should be rigorously dried.

2   The sodium methoxide solution is prepared from freshly cut sodium and "super dry" methanol.

3 Final drying is at oil pump vacuum; the product becomes brown on standing if drying has been insufficient.
4 The ester is recrystallized from 95% ethanol; oxygen-18 is determined by mass spectrometry and purity by isotope dilution analysis using p-phenylphenacyl acetate-1-[14]C.

References

[1]C. R. Hutchinson and C. T. Mabuni, "A Convenient Synthesis of Sodium Acetate-[18]O$_2$," *J. Labelled Compds.*, **13**, 571-574 (1977).

[13]C-BROMOACETIC ACID

Method I

$$CH_3*COOH \xrightarrow[\text{2. } H_2O]{\text{1. } (CF_3CO)_2O, \, Br_2, \, PBr_3} BrCH_2*COOH$$

Procedure[1]

Bromoacetic-1-^{13}C acid Trifluoroacetic anhydride (44.9 g, 0.214 mol) is carefully added, with cooling, to acetic-1-^{13}C acid (5.50 g, 90.4 mmol) that contains water (0.7 g, 39 mmol) (Note 1). Phosphorus tribromide (0.40 g, 1.5 mmol) is added to the mixture, and the reaction flask is heated in an oil bath at 60°. Bromine (14.4 g, 90.2 mmol) is slowly added to the refluxing mixture at a rate such that a pale bromine color is just maintained. After addition is complete (6-7 hr) the mixture is cooled, water (3.39 g, 0.188 mol) is carefully added (Note 2), and trifluoracetic acid and hydrogen bromide are removed by distillation (Note 3). The yield of bromoacetic-1-^{13}C acid, mp 46-48°, is 11.8 g (95%) (Note 4).

Method II

$$*CH_3COOH \xrightarrow[\text{2. } H_2O]{\text{1. } P, \, Br_2} Br*CH_2COOH$$

Procedure[2]

Bromoacetic-2-^{13}C acid Acetic-2-^{13}C acid (42.0 g, 0.690 mol) and red phosphorus (7.0 g, 0.23 mol) in a 500-ml 3-neck flask fitted with a condenser with calcium chloride drying tube, dropping funnel, and magnetic stirrer is cooled in an ice bath. Bromine (147 g, 0.91 mol) is added slowly from the dropping funnel (Note 5). The mixture is then refluxed overnight, cooled, and treated with methylene chloride (150 ml). Most of the solvent (approx. 100 ml) is distilled, and water (13 ml, 0.72 mol) is added dropwise. Additional methylene chloride (150 ml) is added and distilled. Distillation at reduced pressure gives the product, bp 63-66° at 4 mm Hg (91.9 g, 95% yield).

Notes

1 The reaction is very exothermic owing to reaction of the water as well as formation of the mixed anhydride. It is recommended that the addition be made very
   slowly with adequate cooling and stirring in apparatus having a Dry Ice condenser and protected from atmospheric moisture.

2 The reaction is extremely vigorous. The amount of water used represents only a
   small excess over that required.

3 A small amount of product is recovered by evaporation of the distillate in a
   stream of nitrogen.

4 With unlabeled anhydrous acetic acid a quantitative yield has been obtained.

5 The bromine is added slowly at first because of vigorous reaction with phosphorus; addition is complete in a few hours.

Other Preparations

Bromoacetic-2-^{13}C acid has been prepared from the acetic acid using bromine with
phosphorus trichloride and no solvent; a very pure product is obtained by sublimation at $50°$ and 0.1 mm Hg.[3] Acetic anhydride is effective for bromination, but
must be labeled to avoid dilution of isotopic concentration.[4] Both Methods I and II
are also applicable to the other isotope isomers, including bromoacetic-^{13}C$_2$ acid.
(See also Malonic-1-^{13}C Acid.)

References

[1]J. L. Roberts and C. D. Poulter, "2', 3', 5'-Tri-*O*-benzoyl[4-^{13}C]uridine. An Efficient, Regio-
specific Synthesis of the Pyrimidine Ring," *J. Org. Chem.*, 43, 1547-1550 (1978).

[2]V. N. Kerr, in "Annual Report of the Biomedical and Environmental Research Program of the
LASL Health Division," compiled by E. C. Anderson and E. M. Sullivan, Los Alamos Scientific
Laboratory Report LA-5883-PR, February 1975, p. 125.

[3]I. J. G. Climie and D. A. Evans, "An Efficient Synthesis of [2-^{13}C]-Monobromoacetic Acid,"
J. Labelled Compds., 13, 311-313 (1977).

[4]D. G. Ott, C. T. Gregg, V. N. Kerr, V. H. Kollman, and T. W. Whaley, "Organic Synthesis and
Biosynthesis of ^{13}C-Labelled Compounds for Clinical Investigations," *Proceedings of a Sympo-
sium on Radiopharmaceuticals and Labelled Compounds*, International Atomic Energy Agency,
Vienna, 267-279 (1973).

ETHYL CHLOROACETATE-1-^{13}C

$$CH_3{*}COOH \xrightarrow[110°]{Cl_2,\ Ac_2O} ClCH_2{*}COOH$$

$$\xrightarrow[TsOH,\ CHCl_3]{EtOH} ClCH_2{*}COOEt$$

Procedure[1]

A mixture of acetic-1-^{13}C acid (29 g, 0.48 mol) and acetic anhydride (6 g, 0.059 mol) (Note 1) is heated in a bath at 110°, and chlorine is introduced for 17 hr (Note 2). On cooling the chloroacetic-1-^{13}C acid crystallizes (41 g, 73% yield). The crude acid is heated on a water bath with ethanol (50 g, 1.1 mol), toluenesulfonic acid (2.5 g), and chloroform (50 ml) until water ceases to form (12 hr) (Note 3). The solution is cooled, washed with water, bicarbonate solution, again with water, and fractionally distilled to give the product, bp 136-144° (47.5 g, 88% yield) (Note 4).

Notes

1 Use of acetic anhydride lowers the carbon-13 concentration; for certain purposes this is acceptable. When isotopic dilution is undesirable, labeled anhydride can be used, or trifluoracetic anhydride (see ^{13}C-Bromoacetic Acid, Method I).
2 Condenser temperature is maintained at −5° so that acetyl chloride, which is an intermediate, returns to the reaction mixture.
3 Presumably a Dean-Starke type of trap is used.
4 The overall yield of ester from acetic acid is 65%.

References

[1]B. Stuetzel, W. Ritter, and K-F. Elgert, "^{13}C-Enriched Monomers for NMR-Spectroscopy of Polymers," *Angew. Makromol. Chem.*, 50, 21-41 (1976).

CYANOACETIC-1-^{13}C ACID

$$BrCH_2*COOH \xrightarrow[\substack{2.\ KCN \\ 3.\ HCl}]{1.\ H_2O,\ K_2CO_3} NCCH_2*COOH$$

Procedure[1]

A stirred solution of bromoacetic-1-^{13}C acid (2.78 g, 20 mmol) in water (6 ml) is neutralized with potassium carbonate (1.49 g, 11 mmol). Potassium cyanide (1.40 g, 21.5 mmol) in water (5 ml) is added slowly, and the mixture is stirred at ambient temperature for 15 min, then at 60° for 25 min. The solution is cooled to 10°, acidified with hydrochloric acid (6 *N*, 3.63 ml, 21.5 mmol), and extracted with ether in a continuous extractor for 72 hr. The ether extract is dried over sodium sulfate and evaporated to afford the product (1.71 g, 100% yield).

References

[1]J. L. Roberts and C. D. Poulter, "2',3';5'-Tri-*O*-benzoyl[4-^{13}C]uridine. An Efficient, Regiospecific Synthesis of the Pyrimidine Ring," *J. Org. Chem.*, 43, 1547-1550 (1978).

PHENYLACETIC-$^{13}C_2$ ACID

$$Ph*COOH \xrightarrow[Et_2O]{LiAlH_4} Ph*CH_2OH \xrightarrow[C_6H_6]{SOCl_2}$$

$$Ph*CH_2Cl \xrightarrow[\substack{2.\ *CO_2 \\ 3.\ HCl}]{1.\ Mg,\ Et_2O} Ph*CH_2*COOH$$

Procedure[1]

Benzoic-^{13}C acid (7.2 g, 59 mmol) is reduced by lithium aluminum hydride (2.8 g, 74 mmole) in ether (220 ml) to give benzyl-α-^{13}C alcohol (5.2 g, 81% yield). The alcohol is refluxed with thionyl chloride (12.5 g, 0.105 mol) for 4 hr, benzene is added, and the benzene and excess thionyl chloride are distilled. Distillation affords benzyl-α-^{13}C chloride, bp 120° at 60 mm Hg (5.0 g, 81% yield). The chloride (39 mmol) is converted to the Grignard reagent using magnesium (1.5 g, 62 mmol) and ether (110 ml), and carbonated at −30° with the carbon-^{13}C dioxide generated from barium carbonate-^{13}C (8.1 g, 41 mmol) and sulfuric acid (conc., 60 ml) (Note 1). Acidification of the reaction mixture, ether extraction, conversion to the sodium salt, reextraction, acidification, and filtration afford the product (4.5 g, 81% yield) (Note 2).

Notes

1 A procedure is described for generating and introducing the carbon dioxide in a closed, evacuated system.
2 Recrystallization from water gives mp 75-77° (4.0 g).

References

[1]M. Pomerantz and R. Fink, "Synthesis of 1-[(^{13}C)Methyl]-3-phenyl(1,3-$^{13}C_2$)bicyclo[1.1.0]-butane-exo,exo-2,4-dicarboxylate," J. Labelled Compds., 16, 275-286 (1979).

N-(5-FLUORO-2,4-DINITROPHENYL)ACETAMIDE-1-^{13}C

$$CH_3*COONa + Me_3COCl \xrightarrow{MeCN} CH_3*COOCOMe_3$$

Procedure[1] (Note 1)

A suspension of finely crystalline, anhydrous sodium acetate-1-^{13}C (0.45 g, 5.5 mmol) in acetonitrile (20 ml) is combined with pivaloyl chloride (0.67 ml, 5.5 mmol) and refluxed with stirring for 1-3 hr. The mixture is cooled, filtered to remove sodium chloride, and evaporated at reduced pressure. The residual acetic-1-

[13]C pivalic anhydride is combined with 5-fluoro-2,4-dinitroaniline (1.0 g, 5 mmol), acetonitrile (0.5 ml), and sulfuric acid (2 microdrops) (Note 2), and the solution is heated on a steam bath with stirring for 3 hr. Water (10 ml) is added and the mixture is neutralized with sodium bicarbonate (4% solution), chilled, and filtered. The solid is washed with water, dried, and recrystallized from benzene-hexane to give the product, mp 102-113° (0.87 g, 72% yield).

Notes

1 The mixed acetic pivalic anhydride provides a convenient, efficient, and generally applicable synthesis of acetanilides.
2 Removal of sodium chloride and use of sulfuric acid as a catalyst is not required with other aniline derivatives.

Other Preparations

By similar procedures *N*-(4-cyanophenyl)acetamide-1-[13]C (76% yield), *N*-(4-formylphenyl)acetamide-1-[13]C (53% yield), *N*-(3-carboxyphenyl)acetamide-1-[13]C (57% yield), and *N*-(2-carboxyphenyl)acetamide-1-[13]C (56% yield) were prepared.

References

[1]W. J. Horton, S. P. Adams, and M. W. Niermayer, "Acetylation with Acetic-1-[13]C Pivalic Anhydride," *J. Labelled Compds.*, **13**, 611-617 (1977).

N-(1,1,3,3-TETRAMETHYLBUTYL)ACETAMIDE-2-[13]C (*N-t*-OCTYLACETAMIDE-2-^{13}C)

$$*CH_3COCl + H_2NCMe_2CH_2CMe_3 \xrightarrow{C_5H_5N} *CH_3CONHCMe_2CH_2CMe_3$$

Procedure[1]

To *t*-octylamine (1.5 ml, 11 mmol) in pyridine (1.5 ml) is added rapidly acetyl-2-^{13}C chloride (0.450 ml, 6.4 mmol) (Note 1). The mixture is stirred for 18 hr at room temperature, acidified with dilute phosphoric acid, and extracted with ether. Evaporation of the ether affords the product, mp 99° (999 mg, 90% yield) (Note 2).

Notes

1 A white precipitate forms immediately.
2 Recrystallization from petroleum ether gives mp 100.5°. ^{1}H-NMR data are given.

Other Preparations

By the same procedure *N-t*-octylbenzamide-1-^{13}C, *N-t*-butylbenzamide-1-^{13}C, and *N*-(phenyl-d$_5$)benzamide-1-^{13}C are also prepared.

References

[1]G. Chapelet-Letourneux and A. Rassat, "Synthesis of Amines and Nitroxide Radicals Labeled with Carbon-13 and Deuterium," *Bull. Soc. Chim. Fr.*, **1971**, 3216-3221.

ACETONITRILE-2-^{13}C

Method I

$$*CH_3I \xrightarrow[\text{HOCH(CH}_2\text{OH)}_2]{\text{KCN}} *CH_3CN$$

Procedure[1]

Methyl-^{13}C iodide (40.0 g, 0.280 mol) is mixed with dry glycerol (25 ml), and potassium cyanide (18.2 g, 0.330 mol) (Note 1) is added. The heterogeneous system, at 25°, is stirred for 4 hr (Note 2) and then heated in an oil bath to a final temperature of 180° to distill the product (10.3 g, 90% yield).

Method II

$$*CH_3I \xrightarrow[\text{DMSO}]{\text{NaCN}} *CH_3CN$$

Procedure[2]

A solution of methyl-^{13}C iodide (182.8 g, 1.28 mol) in dimethyl sulfoxide (200 ml) is added over 2.5 hr (Note 3) to a stirred mixture of sodium cyanide (75.3 g, 1.54 mol) and dimethyl sulfoxide (250 ml). After the reaction has been stirred an additional 30 min, the product is distilled directly from the mixture (51.2 g, 95% yield).

Notes

1 The weight does not agree with the moles given;[1] 18.2 g is 0.28 mol.
2 The reaction is exothermic.
3 The addition rate is controlled such that the mixture remains at ambient temperature.

References

[1]D. E. Klinge and H. C. van der Plas, "Didehydrohetarenes (XXXVII). On the Existence of 4,5-Didehydropyridazine," *Rec. Trav. Chim.*, **95**, 34-36 (1976).
[2]T. W. Whaley, G. H. Daub, V. N. Kerr, T. A. Lyle, and E. S. Olson, "Syntheses with Stable Isotopes: DL-Valine-^{13}C$_3$," *J. Labelled Compds.*, **16**, 809-817 (1979).

GLYCINE-2-^{13}C

$$Br*CH_2COOH \xrightarrow{NH_4OH} H_2N*CH_2COOH$$

Procedure[1]

Bromoacetic-2-^{13}C acid (55.0 g, 0.393 mol) in water (560 ml) (Note 1) is added with vigorous stirring over 6.5 hr to ammonium hydroxide (conc., 8.5 liter, 125 mol) in a 12-liter flask chilled in an ice bath (Note 1). The solution is transferred to a large evaporating dish and evaporated to a small volume in the draft of a hood. The concentrated solution (approx. 500 ml) is treated with Norit, filtered, and evaporated at reduced pressure until precipitation begins. The solution is heated, and methanol is added at the boiling point until crystallization starts. After cooling overnight, filtration affords the product (23.5 g, 79% yield). From the filtrate a second crop is obtained (3.35 g, 90% total yield) (Note 2).

Notes

1 Dissolving the acid in water and chilling the ammonium hydroxide solution help prevent excessive reaction at the tip of the dropping funnel. The large excess of ammonia prevents formation of significant quantities of iminodi(acetic-2-^{13}C) acid and nitrilotri(acetic-2-^{13}C) acid.

2 Since ammonium bromide is quite soluble in methanol, separation from glycine is considerably simpler than if the chloroacetic acid had been used. The crystallization process effectively eliminates small amounts of iminodiacetic and nitrilo-triacetic acids also formed. Evaporation readily converts ammonium glycinate to ammonia and glycine.

Other Preparations

The method has also been used to prepare glycine-1-^{13}C and glycine-^{13}C$_2$.[1] The three carbon-13 isotope isomers have also been obtained by the Gabriel synthesis;[2] however, the ammonolysis procedure has been found to be more efficient since there are fewer steps and the acid, rather than the ester, is the starting material. Glycine-1-^{13}C and -^{13}C$_2$ have been converted to the respective ^{13}C-glycocholic acids[3] using a published procedure.[4]

References

[1] V. N. Kerr and D. G. Ott, in "Annual Report of the Biomedical and Environmental Research Program of the LASL Health Division," compiled by E. C. Anderson and E. M. Sullivan, Los Alamos Scientific Laboratory Report LA-5883-PR, February 1976, p. 130.

[2] V. N. Kerr and D. G. Ott, in "Annual Report of the Biological and Medical Research Group (H-4)," C. R. Richmond and G. L. Voelz, Eds., Los Alamos Scientific Laboratory Report LA-4923-PR, April 1976, pp. 122-123.

[3]N. W. Solomons, D. A. Schoeller, J. B. Wagonfeld, D. Ott, I. H. Rosenberg, and P. D. Klein, "Application of a Stable Isotope (^{13}C)-Labeled Glycocholate Breath Test to Diagnosis of Bacterial Overgrowth and Ileal Dysfunction," *J. Lab. Clin. Med.*, **90**, 431-439 (1977).

[4]L. Lack, F. O. Dorrity, Jr., T. Walker, and G. D. Singletary, "Synthesis of conjugated bile acids by means of a peptide coupling agent," *J. Lipid Res.*, **14**, 367-370 (1973).

GLYCINE-^{13}C$_2$-^{15}N

$$\text{(phthalimide)}^*NK + Br^*CH_2^*COOEt \xrightarrow[\text{2. HCl, HOAc}]{\text{1. DMF}}{} H_2^*N^*CH_2^*COOH$$
$$\text{3. NH}_4\text{OH}$$

Procedure[1]

Ethyl bromoacetate-^{13}C$_2$ (5.06 g, 30.1 mmol), dimethyl formamide (21 ml), and potassium phthalimide-^{15}N (5.59 g, 30.1 mmol) are stirred for 3 days at room temperature. The mixture is treated with water (26 ml) and extracted with chloroform (25 ml, followed by 2 × 5 ml). The extract is washed with sodium hydroxide (0.1 N, 10 ml) and water (4 × 10 ml), dried over calcium sulfate, filtered, and partially evaporated. To the residue are added acetic acid (18 ml), hydrochloric acid (conc., 18 ml), and water (18 ml), and the mixture is refluxed for a day (Note 1). After cooling overnight, the mixture is filtered to remove phthalic acid (4.26 g, 26 mmol) and evaporated at reduced pressure to a syrup. Water, ethanol-water, and benzene are successively added and evaporated to remove acetic and hydrochloric acids; the procedure was repeated six times. The solid residue is taken up in water, and the pH of the solution is adjusted to 5.7 with ammonium hydroxide. Water is added to a total volume of 20 ml, and 2-propanol (140 ml) is added dropwise with stirring (Note 2). After cooling overnight, the solid (1.56 g) is filtered, washed, dissolved in water (6 ml), and precipitated by slow addition of methanol (42 ml). The product is filtered, washed with aqueous methanol and methanol, and dried (1.51 g, 65% yield).

Notes

1 The residual chloroform is allowed to escape as reflux is started.

2 The ratio of 2-propanol to water of 7:1 allows precipitation of the glycine while leaving the ammonium chloride in solution. Occasionally, chloride also precipitates, and the process must be repeated. It has subsequently been found that hydrolysis of the phthalimido ester with hydrobromic acid, rather than hydrochloric acid, eliminates this problem because ammonium bromide is quite soluble in the lower alcohols.

Other Preparations

Glycine-2-^{13}C-^{15}N has been obtained by the same procedure starting from ethyl bromoacetate-2-^{13}C.

References

[1]D. G. Ott and V. N. Kerr, in "Annual Report of the Biological and Medical Research Group (H-4)," C. R. Richmond and G. L. Voelz, Eds., Los Alamos Scientific Laboratory Report LA-4923-PR, April 1976, p. 107.

ETHYL N-ACETYLGLYCINATE-^{15}N

$$H_2*NCH_2COOH \xrightarrow[H_2O]{Ac_2O} Ac*NHCH_2COOH$$

$$\xrightarrow[HCl]{EtOH} Ac*NHCH_2COOEt$$

Procedure[1]

N-Acetylglycine-^{15}N   To a solution of glycine-^{15}N (7.0 g, 92 mmol) in water (30 ml) is added acetic anhydride (10.0 ml, 106 mmol) in one portion. The mixture is stirred for 20 min at ambient temperature and chilled for 20 hr. The precipitate is collected, washed with water (5 ml), and dried to give the product as white crystals, mp 203-204° (6.6 g). Concentration of the filtrate gives a second crop (0.90 g) (total yield 7.5 g, 69%).

Ethyl N-acetylglycinate-^{15}N   To ice-cold hydrogen chloride (3%) in ethanol (300 ml) is added N-acetylglycine-^{15}N (15.3 g, 0.13 mol). The mixture is stirred for 2 hr at ambient temperature and refrigerated for 20 hr. The solvent is removed at reduced pressure, and ethanol (50 ml), followed by two portions of benzene (50 ml), are successively added and evaporated. The residue is dissolved in dichloromethane (300 ml) and stirred with anhydrous potassium carbonate (15 g) for 1 hr. After filtration and removal of solvent, the residue is distilled to afford the product, bp 100-105° at 0.25 mm Hg (15.9 g, 84% yield).

References

[1]C. SooHoo, J. A. Lawson, and J. I, DeGraw, "Synthesis of Multilabeled Histidine," J. Labelled Compds., 13, 97-102 (1977).

CREATINE-1-^{13}C
[(1-METHYLGUANIDINO)ACETIC-1-^{13}C ACID]

$$H_2NCH_2*COOH \xrightarrow[2.\ H_2,\ Pd/C]{1.\ PhCH_2Cl} PhCH_2NHCH_2*COOH$$

$$\xrightarrow[2.\ H_2,\ Pd/C]{1.\ HCHO,\ HCOOH} CH_3NHCH_2*COOH$$

$$\xrightarrow[2.\ CH_3SC(NH)_2^+\ I^-]{1.\ NaOH} H_2NC(NH)NMeCH_2*COOH$$

Procedure[1]

Sarcosine-1-^{13}C Benzyl chloride (4.0 ml, 35 mmole) is added over 10 min to a re-fluxing solution of glycine-1-^{13}C (1.20 g, 16 mmol), potassium hydroxide (85%, 3.2 g, 48 mmol), water (9 ml), and ethanol (9 ml). After the mixture is heated 1 hr, it is cooled, evaporated to half volume, and treated with an equal volume of methylene chloride and water (20 ml). The layers are separated, the aqueous phase is extracted further with methylene chloride (2 × 10 ml), and the combined methy-lene chloride extracts are washed with water (10 ml). The aqueous layers, acidified to pH 6 with acetic acid, afford N,N-dibenzylglycine-1-^{13}C, mp 193-195° (3.46 g, 85% yield).

The dibenzylglycine (3.32 g, 13.0 mmol) in methanol (39 ml), hydrochloric acid (conc., 1.63 ml, 19.5 mmol), and water (2.0 ml) is hydrogenated over prere-duced 10% palladium on charcoal (1.30 g) at atmospheric pressure and 22°. The mixture is filtered and evaporated to give N-benzylglycine-1-^{13}C hydrochloride, mp. 215.5-217.5° (2.45 g, 93% yield). Passage through a column of Bio-Rad AG3-X4 resin and evaporation of the water (500 ml) at reduced pressure afford N-benzyl-glycine-1-^{13}C, mp 195-197.5° (1.94 g, 97% recovery). A mixture of the benzyl-glycine (1.94 g, 11.8 mmol), formic acid (2.8 ml, 71 mmol), and aqueous formalde-hyde (36%, 2.1 ml, 18 mmol) is heated on a steam bath for 15 min (Note 1), cooled, and evaporated at reduced pressure to give a white, slightly moist solid (2.22 g). The crude product is dissolved in boiling alcohol (18 ml), and ether (108 ml) is added dropwise with vigorous stirring. The product is filtered, washed with ether and hexane, and dried to give N-benzyl-N-methylglycine-1-^{13}C, mp 188.5-191.5° (1.87 g, 89% yield).

The above product (1.87 g, 10.5 mmol) with prereduced 10% palladium on charcoal (1.05 g) in aqueous acetic acid (90%, 50 ml) is hydrogenated for 2 hr at 22° and atmospheric pressure to afford sarcosine-1-^{13}C (0.98 g) (Note 2).

Creatine-1-^{13}C The sarcosine above (10.5 mmol) in water (1.5 ml) is neutralized with sodium hydroxide (10 N, 1.05 ml), the yellow solution is treated with S-methylthiuronium iodide (2.29 g, 10.5 mmol) in water (2.0 ml) at 30° with vigor-ous stirring over 2.5 hr, and stirred an additional 12 hr. The precipitate is collected by filtration and dried to give a yellow solid (0.92 g). Concentration of the filtrate to 3 ml and treatment with ethanol (12 ml) give an additional 0.10 g. A solution of the crude product (1.02 g) in warm water (75 ml) is filtered, concentrated to 40 ml, treated with ethanol (160 ml), and cooled at 4° overnight to give the anhydrous product as lustrous plates (0.90 g, 66% yield). Recovery of starting material from the filtrates and retreatment give additional product as the hydrate (1.1 mmol) (Note 3).

Notes

1 Carbon dioxide evolution ceases after this time.
2 The pale-yellow product, in apparent yield of 105%, is used without further purification.

3   The total yield based on *N*-benzylsarcosine is 76%.

References

[1]G. L. Rowley and G. L. Kenyon, "The Conversion of Isotopically Labeled Glycine to 1-Methyl-2-amino-2-imidazolin-4-one (Creatinine)," *J. Heterocycl. Chem.*, 9, 203-205 (1972).

$^{15}N_1$-CREATINE
[(1-METHYLGUANIDINO-1-^{15}N)ACETIC ACID]

$$H_2*NCH_2COOH \xrightarrow[\text{NaOH}]{\text{PhSO}_2\text{Cl}} PhSO_2*NHCH_2COOH$$

$$\xrightarrow[\text{NaOH}]{\text{Me}_2\text{SO}_4} PhSO_2*NMeCH_2COOH \xrightarrow[\text{2. IR120(H}^+)]{\text{1. H}_2\text{SO}_4} Me*NHCH_2COOH$$

$$\xrightarrow[\text{NH}_4\text{OH, NaCl}]{\text{H}_2\text{NCN}} H_2NC(NH)*NMeCH_2COOH$$

Procedure[1]

Sarcosine-^{15}N   To a vigorously stirred solution of glycine-^{15}N (67 mmol) and sodium hydroxide (1 *M*, 75 ml) in a 250-ml beaker is added benzenesulfonyl chloride (86 mmol). During the subsequent 1-hr reaction time additional sodium hydroxide (3 *M*, 30 ml) is added in portions to keep the reaction mixtue alkaline. The solution is filtered, acidified with hydrochloric acid (conc., 15 ml), and allowed to stand for 3 hr at 4°. The precipitated *N*-(benzenesulfonyl)glycine-^{15}N is filtered and dried under reduced pressure (64 mmol, 96% yield). The sulfonamide is dissolved in sodium hydroxide solution (3 *M*, 65 ml), filtered, and treated with dimethyl sulfate (130 mmol) in six equal portions during a period of 1 hr. Cold hydrochloric acid (conc., 10 ml) is added, and the solution is allowed to stand for 3 hr at 4°. The precipitated *N*-benzenesulfonylsarcosine-^{15}N is filtered, dried (63 mmol, 98% yield), and hydrolyzed by boiling 5 hr in sulfuric acid (7 *M*). The sarcosine-^{15}N is isolated by column chromatography on Amberlite IR120(H$^+$) and recrystallized from aqueous ethanol (52 mmol, 83% yield).

Creatine-^{15}N   A solution of sarcosine-^{15}N (52 mmol) and sodium chloride (52 mmol) in water (10 ml) is treated with cyanamide (98 mmol) in water (2.5 ml) and ammonium hydroxide (conc., 0.3 ml) (Note 1). After 2 days at room temperature the precipitated product is filtered and dried under vacuum (45 mmol, 87% yield). Recrystallization from water (50 ml), with filtration of the hot solution, gives the purified product as the monohydrate (40 mmol, 89% recovery) (Note 2).

Notes

1   Addition of ammonia catalyzes the reaction. The sodium chloride increases the yield, perhaps by salting out the product.

2   The product is greater than 99% pure by gas-liquid chromatographic analysis; assay for creatinine shows less than 0.01%. The overall yield from glycine-^{15}N is 59%.

References

[1] W. Greenaway and F. R. Whatley, "A Convenient Synthesis of Creatine-^{15}N from Glycine-^{15}N via Sarcosine-^{15}N," *J. Labelled Compds.*, 14, 611-615 (1978).

OXALIC-^{13}C$_2$ ACID

$$\text{H*COONa} \xrightarrow[\text{2. HCl}]{\text{1. Na}_2\text{CO}_3, \, 360°} \text{HOO*C*COOH}$$

Procedure[1]

Sodium formate-^{13}C (10.2 g, 0.15 mol) is ground with sodium carbonate (31.8 g, 0.30 mol), placed in an open-end glass tube, and pyrolyzed at 360° for 30 min. After cooling, the contents of the tube are transferred to a 500-ml beaker, using a minimum amount of water, and acidified with conc. hydrochloric acid to pH 1. Filtration gives the product (5.1 g); evaporation of the filtrate and sublimation of the residue at 200° and 0.07 mm Hg gives additional product, mp 99-100° dec. (1.2 g) (total yield 6.3 g, 93%) (Note 1).

Notes

1   Purity of the product is established by infrared and mass spectrometry.

References

[1] B. D. Andresen, "Synthesis of Sodium Formate-^{13}C and Oxalic Acid-^{13}C$_2$," *J. Org. Chem.*, 42, 2790 (1977).

PROPIONIC-2,3-^{13}C$_2$ ACID

$$\text{*CH}_3\text{*CH}_2\text{OH} + \text{CO} \xrightarrow[\text{H}_2\text{O}, \, 200°]{\text{RhCl}_3, \, \text{HI}} \text{*CH}_3\text{*CH}_2\text{COOH}$$

Procedure[1] (Note 1)

Ethanol-^{13}C$_2$ (22.9 g, 0.48 mol), water (30 ml), hydriodic acid (57%, 10 ml), and rhodium trichloride (0.8 g) are placed in a 1-liter stirred Hastelloy C autoclave. Carbon monoxide (500 psi, 1.5 mol) is admitted, the temperature is brought to 200°, and the reaction is allowed to proceed until the pressure ceases to decrease (approx. 10 hr; pressure decrease of about 290 psi, 0.52 mol). The mixture is cooled, withdrawn from the autoclave, treated with phosphorous acid (0.5 ml) (Note 2), and distilled. The resulting aqueous solution of the product is redistilled after addition of silver sulfate (Note 2), the product is neutralized by titration with sodium hydroxide, and the solution is evaporated to give sodium propionate-2,3-^{13}C$_2$ (41.1 g, 88% yield). Anhydrous propionic-2,3-^{13}C$_2$ acid is obtained by treating the salt with excess phosphoric acid (85%) and subsequent bulb-to-bulb vacuum transfer (Note 3).

Notes

1 Carbonylation of ethanol-1-^{13}C or ethanol-2-^{13}C by this method produces propionic-1,2-^{13}C$_1$ acid; that is, the isotopic position is scrambled. Presumably a symmetrical intermediate, such as ethylene, is involved. Similarly, carbonylation of either 1-propanol or 2-propanol yields a mixture of butyric and isobutyric acids. Application of this method as a general preparation of labeled carboxylic acids should be made only after consideration of such isomerization possibilities.

2 Phosphorous acid is added to prevent formation and distillation of iodine, and silver sulfate immobilizes the iodide.

3 Physical properties and NMR and IR spectroscopic data are given.

Other Preparations

Sodium propionate-1-^{13}C has been prepared using the same procedure by carbonylation of ethanol, but with a limited amount of carbon-^{13}C monoxide.[2] The resulting ester fraction is hydrolyzed with sodium hydroxide, combined with the acid fraction, neutralized with sodium hydroxide, and evaporated. The salt is isolated in approx. 90% yield following recrystallization from ethanol-water.

References

[1] V. N. Kerr and D. G. Ott, "Preparation of D- and L-Alanine-2,3-^{13}C$_2$," *J. Labelled Compds.*, 15, 503-509 (1978).

[2] D. G. Ott, "Carbon-13 Labelled Compounds," *Kagaku No Ryoiki Zokan*, 107, 95 (1975).

2-BROMOPROPIONIC-2,3-^{13}C$_2$ ACID

$$*CH_3*CH_2COOH \xrightarrow[\text{2. } H_2O]{\text{1. } Br_2, PBr_3, CCl_4} *CH_3*CHBrCOOH$$

Procedure[1]

Propionic-2,3-^{13}C$_2$ acid (29.6 g, 0.39 mol) (Note 1) in carbon tetrachloride is treated with phosphorus tribromide (13.5 ml, 0.07 mol) and heated to reflux. A solution of bromine (81 g, 0.51 mol) in carbon tetrachloride (20 ml) is added over an 11-hr period, and the reaction mixture is refluxed for an additional 12 hr. After the colorless solution is cooled, water (4 ml) is added, and the mixture is refluxed for 20 min. Following distillation of the carbon tetrachloride, the product is distilled at reduced pressure, bp 116-118° at 20 mm Hg (57.0 g, 94% yield) (Note 2).

Notes

1 Aqueous propionic acid can be dried by azeotropic distillation with carbon tetrachloride and the solution used directly.
2 NMR and IR data are given.

References

[1] V. N. Kerr and D. G. Ott, "Preparation of D- and L-Alanine-2,3-$^{13}C_2$," *J. Labelled Compds.*, **15**, 503-509 (1978).

METHYL PROPIOLATE-1-^{13}C

$$*CO_2 + HC{\equiv}CMgBr \xrightarrow[\text{2. }H^+]{\text{1. THF}} HC{\equiv}C*COOH \xrightarrow[\text{Et}_2O]{\text{CH}_2N_2} HC{\equiv}C*COOMe$$

Procedure[1]

A solution of the Grignard reagent prepared from magnesium (1.8 g, 74 mmol) and ethyl bromide (8.9 g, 82 mmol) in tetrahydrofuran (50 ml) is added dropwise to a saturated solution of acetylene in tetrahydrofuran (100 ml) at room temperature through which acetylene is continuously bubbled (Note 1). The ethynylmagnesium bromide solution at 10° is transferred dropwise to a solution of carbon-^{13}C dioxide, from barium carbonate-^{13}C (12.02 g, 60.7 mmol), in tetrahydrofuran (25 ml) at −196°. The resulting thick, white suspension is warmed to 25°, stirred an additional hour, and extracted with acidified saturated ammonium sulfate solution (3 × 60 ml) (Note 2). The ether solution is dried over sodium sulfate and the solvents are fractionated to give a concentrated tetrahydrofuran solution (50% by weight) of propiolic-1-^{13}C acid (3.29 g, 77% yield). Addition of an equimolar quantity of diazomethane in ether to an ice-cold ether solution of the acid gives the ester (3.04 g, 75% yield from the acid) (Note 3).

Notes

1 Lithium acetylide, from acetylene and butyllithium, is also satisfactory; because it is formed quantitatively, it is the preferred reagent when acetylene-$^{13}C_2$ is used in preparation of 5-fluoro-2′-deoxyuridine-5,6-$^{13}C_2$.
2 Addition of ether, up to 200 ml, may be used to break emulsions.

3   An excess of diazomethane must be avoided owing to further reaction to form a pyridazole. Fractional distillation of the ester greatly reduced the yield; therefore the solution is used directly after concentrating to a small volume.

References

[1]W. H. Dawson and R. B. Dunlap, "An Improved Synthesis of 5-Fluoro-2′-deoxyuridine Incorporating Isotopic Labels," *J. Labelled Compds.*, **16**, 335-343 (1979).

ZINC LACTATE-2,3-$^{13}C_2$

$$*CH_3*CHBrCOOH \xrightarrow[H_2O]{ZnO} (*CH_3*CHOHCOO)_2Zn$$

Procedure[1]

2-Bromopropionic-2,3-$^{13}C_2$ acid (7.74 g, 50 mmol) is dissolved in a slurry of zinc oxide (25 g, 0.25 mol) in water (80 ml). The mixture is refluxed for 9 hr, cooled, filtered, and evaporated at reduced pressure to give the product (5.93 g, 96% yield) (Note 1).

Notes

1   Solutions of DL-lactic-2,3-$^{13}C_2$ acid or its sodium salt can be obtained by treatment of a solution of the zinc salt with Dowex-50(H^+) or Dowex-50(Na^+), respectively.

References

[1]D. G. Ott, "Carbon-13 Labelled Compounds," *Kagaku No Ryoiki Zokan*, **107**, 95 (1975).

ZINC L-LACTATE-$^{13}C_3$

Glucose-$^{13}C_6$,
Fructose-$^{13}C_6$, and/or $\xrightarrow{\begin{array}{l}\text{1. Fermentation}\\\text{2. } H_2SO_4,\ Et_2O\\\text{3. } ZnCO_3\end{array}}$
Sucrose-$^{13}C_{12}$

$$(L\text{-}*CH_3*CHOH*COO)_2Zn$$

Procedure[1] (Note 1)

A mixture of glucose-$^{13}C_6$ (0.13 mol), fructose-$^{13}C_6$ (0.12 mol), and sucrose-$^{13}C_{12}$ (0.24 mol) (Note 2) is combined with the fermentation medium that is prepared by dissolving trypton (5 g), yeast extract (5 g), Tween-80 (2 ml), L-cysteine hydrochloride (100 mg), sodium thioglycolate (50 mg), vitamin B_{12} (10 μg), sodium acetate (10 g), and Brin's inorganic salts solution (5 ml) in water (1 liter). The medium is adjusted to pH 6.8 and autoclaved. The inoculum of *Lactobacillus delbrueckii* (ATCC No. 11443) is from 10 ml of inoculated medium that has been kept at 37° for 16 hr. The fermentation is carried out (Note 3) until production of lactic acid ceases (stable pH) and the sugars are depleted (approx. 20 hr). The mixture is cooled to 10°, acidified with cold sulfuric acid (4 N), centrifuged to remove the cells, and concentrated to a semisolid mass by evaporation at 5° under reduced pressure. The residue is extracted several times with ether, and the combined extracts are evaporated to an oil that is taken up in water. The aqueous solution is washed with ether and then treated with zinc carbonate until the pH remains constant at 6.5. Decolorizing charcoal is added and the suspension is filtered. The solution is concentrated and diluted with alcohol to afford the product as the dihydrate from which the anhydrous salt is obtained by heating at 100° under vacuum for 24 hr (Note 4).

Notes

1   The method is a large-scale modification of that previously described.[2]

2   The sugars are obtained from the photosynthetic incubation of mature excised tobacco leaves in the presence of carbon-^{13}C dioxide.[3] The sugars can be separated by column chromatography on Dowex-50(Ba^{+2}) resin; however, because the fermentation is equally effective for the various sugars, either singly or in combination, separation is unnecessary. A related photosynthetic incubation, employing the marine red alga *Gigartina*, produces galactosylglycerol-[13]C_9, which is hydrolyzed to galactose-[13]C_6 and glycerol-[13]C_3.[4]

3   An automatic Micro Ferm Laboratory Fermentor, New Brunswick Scientific Company, is used.

4   Enzymatic analysis of the reaction mixture shows yields greater than 90%; recovery of the zinc salt is somewhat less. Product purity is determined from optical rotation and enzymatic assay (Rapid Lactate Stat-Pack, Calbiochem). The carbon-13 enrichment is essentially the same as that of the sugar substrate.

References

[1]C. T. Gregg and J. Y. Hutson, in "Annual Report of the Biomedical and Environmental Research Program of the LASL Health Division," compiled by E. C. Anderson and E. M. Sullivan, Los Alamos Scientific Laboratory Report LA-5883-PR, February 1975, pp. 131-132.

[2]M. Brin, "L(+) and D(–) Lactic Acids," *Biochem. Prep.*, 3, 61-64 (1953).

[3]V. H. Kollman, J. L. Hanners, J. Y. Hutson, T. W. Whaley, D. G. Ott, and C. T. Gregg, "Large-Scale Photosynthetic Production of Carbon-13 Labeled Sugars: The Tobacco Leaf System," *Biochem. Biophys. Res. Commun.*, 50, 826-831 (1973).

[4]V. H. Kollman, C. T. Gregg, J. L. Hanners, T. W. Whaley, and D. G. Ott, "Large-Scale Photosynthetic Production of Carbon-13 Labeled Sugars," in *Proceedings of the First International Conference on Stable Isotopes in Chemistry, Biology and Medicine*, Argonne National Laboratory, Argonne, IL, May 9-11, 1973 (P. D. Klein and S. V. Peterson, Eds.), USAEC Report CONF-730525 (1973), pp. 30-40.

SODIUM PYRUVATE-$^{13}C_3$

$$(*CH_3*CHOH*COO)_2Zn \xrightarrow[\text{2. BuOH, H}_2\text{SO}_4]{\text{1. Dowex-50(H}^+\text{)}} *CH_3*CHOH*COOBu$$

$$\xrightarrow[\text{2. NaOH}]{\text{1. CrO}_3, \text{H}_2\text{SO}_4, \text{H}_2\text{O}} *CH_3*CHOH*COONa$$

Procedure[1]

Butyl lactate-$^{13}C_3$ A solution of zinc lactate-$^{13}C_3$ (50 mmol) in water (300 ml) is passed through a column of Dowex-50(H$^+$, WX-8) resin (200 ml, 340 meq). The column is washed with water (800 ml), and the aqueous solution is evaporated to a syrup at reduced pressure. The lactic-$^{13}C_3$ acid is transferred to a 500-ml flask fitted with a reflux condenser and Dean-Starke trap. Butanol (100 ml), benzene (100 ml), and sulfuric acid (conc., 1 ml) are added, and the mixture is refluxed for 1 day. After cooling, alizarin indicator is added, and the solution is neutralized with sodium methoxide. The solution is filtered to remove sodium sulfate, and the benzene is evaporated at reduced pressure. Butanol is distilled at atmospheric pressure to 122°, and the residue is distilled under vacuum to give the butyl ester, bp 69-72° at 10 mm Hg (7.5 g).

Sodium pyruvate-$^{13}C_3$ The ester is taken up in ether (80 ml), chromic acid solution (2.2 M, 63 ml) is slowly added over 2.5 hr, and stirring is continued for another hour. The layers are separated, the aqueous phase is extracted with ether (3 × 70 ml), and the combined extracts are washed with water (3 × 180 ml), saturated sodium bicarbonate solution (2 × 100 ml), and water (2 × 125 ml), dried over sodium sulfate, filtered, and evaporated. The residue of butyl pyruvate-$^{13}C_3$ (4.4 g, 62% yield) (Note 1) is taken up in aqueous ethanol (25%, 100 ml) (Note 2) and hydrolyzed by titrating with sodium hydroxide (1 N, 31 ml) to a stable pH of 10.2. The aqueous solution is filtered, extracted with ether (2 × 75 ml), and evaporated at reduced pressure. The residue is dried in vacuum over solid sodium hydroxide to give the product (3.6 g, 32% overall yield) (Note 3).

Notes

1   ^{1}H-NMR shows the oxidation to be essentially complete.
2   The solution is cloudy.
3   Enzymatic assay and ^{1}H-NMR analysis show no measurable lactate contamination.

Other Preparations

Pyruvic-2-^{13}C acid (contaminated with impurities) has been obtained (36% yield) by hydrolysis of pyruvonitrile-2-^{13}C that was prepared by reaction of acetyl-1-^{13}C bromide with cuprous cyanide (88% yield).[2]

References

[1]V. N. Kerr, in "Annual Report of the Biomedical and Environmental Research Program of the LASL Health Division," compiled by E. C. Anderson and E. M. Sullivan, Los Alamos Scientific Laboratory Report LA-5883-Pr, February 1975, pp. 127-128.

[2]A. Heusler and T. Gaumann, "Synthesis of Aromatic and Alicyclic Six-Membered Rings, Labelled with ^{13}C in Positions 1,3,5," J. Labelled Compds., 12, 541-544 (1976).

ALANINE-2,3-^{13}C$_2$

$$*CH_3*CHBrCOOH \xrightarrow{\ NH_4OH\ } DL\text{-}*CH_3*CH(NH_2)COOH$$

$$\xrightarrow[\text{AcOH}]{\ Ac_2O\ } DL\text{-}*CH_3*CH(NHAc)COOH \xrightarrow{\ Acylase\ I\ }$$

$$\begin{array}{c} L\text{-}*CH_3*CH(NH_2)COOH \\ + \\ D\text{-}*CH_3*CH(NHAc)COOH \end{array} \xrightarrow[\text{2. PhNH}_2, \text{EtOH}]{\text{1. HCl, H}_2\text{O}} D\text{-}*CH_3*CH(NH_2)COOH$$

Procedure[1]

DL-Alanine-2,3-^{13}C$_2$ 2-Bromopropionic-2,3-^{13}C$_2$ acid (53.0 g, 0.34 mol) in water (10 ml) is added over 2 hr to vigorously stirred, cold ammonium hydroxide (conc., 2 liters). After 3 days at room temperature the solution is evaporated at reduced pressure to approx. 100 ml, heated to dissolve the solid, and treated with methanol (400 ml). After standing overnight at 5° the product is filtered, washed with cold methanol, and dried. A second crop is obtained from the filtrate to give a total of 23.8 g (71% yield).

DL-N-Acetylalanine-2,3-^{13}C$_2$ DL-Alanine-2,3-^{13}C$_2$ (21.8 g, 0.24 mol) in acetic acid (100 ml) in a 250-ml flask is heated to 108°. The vigorously stirred suspension is allowed to cool to 100°, acetic anhydride (31 ml, 0.33 mol) is added, and the temperature rises to 110°. After 1 min the homogeneous solution is poured into cold water (500 ml) and evaporated to dryness at reduced pressure to give the product, mp 134-136° (32.5 g, 97% yield).

L-Alanine-2,3-^{13}C$_2$ A solution of DL-N-acetylalanine-2,3-^{13}C$_2$ (31.4 g, 0.24 mol) and hog kidney acylase (1.77 g) in ammonium hydroxide (1.5 N, 200 ml) is kept at 38° for 23 hr. The solution is boiled to coagulate the enzyme, treated with Norit and Celite, filtered, and evaporated at reduced pressure to approx. 125 ml. Ethanol (50 ml) is added, and the solution is filtered to remove traces of protein and evaporated at reduced pressure to a thick slurry. Ethanol (300 ml) is added, evaporated to approx. 200 ml, cooled, and filtered to give the crude product (8.2 g), which is recrystallized by dissolving in hot water (55 ml) and adding boiling ethanol (120 ml) (6.8-g first crop, 0.7-g second crop, 70% yield), $[\alpha]_D^{22}$ = 13.86 (9.02% in 6 N HCl) (Note 1).

D-**Alanine-2,3-^{13}C$_2$**   The filtrate from the enzymatic hydrolysis above (0.12 mol) is applied to a Dowex-50(H$^+$) column (370 ml, 444 mmol) and the acetylated amino acid is eluted with water (1 liter) (Note 2). Evaporation at reduced pressure gives D-N-acetylalanine-2,3-^{13}C$_2$(15.3 g, 92% yield), which is recrystallized from acetone (60 ml); mp 127.5-128.5°, $[\alpha]_D^{26}$ = 44.77 (8.98% in 6 N HCl). The acetylated compound (5.3 g, 40 mmol) in hydrochloric acid (2 N, 75 ml) is refluxed for 4 hr and then evaporated to dryness at reduced pressure. The residue is twice evaporated from water and twice from toluene, taken up in water (200 ml), and treated with aniline (4 ml, 43 mmol). The heated solution is filtered, diluted with ethanol (53 ml), cooled, filtered (after several days in the cold), and washed with cold ethanol to give the product (2.65 g, 73% yield), $[\alpha]_D^{22}$ = -13.38 (9.04% in 6 N HCl).

Notes

1   IR and NMR spectral data are given.
2   The resin treatment removes residual L-amino acid.

References

[1]V. N. Kerr and D. G. Ott, "Preparation of D- and L-Alanine-2,3-^{13}C$_2$," *J. Labelled Compds.*, **15**, 503-509 (1978).

3-AMINOPROPIONIC-1-^{13}C ACID
(β-ALANINE-1-^{13}C)

$$NCCH_2*COOH \xrightarrow[\text{2. IR-4B}]{\text{1. } H_2, PtO_2, HCl} H_2NCH_2CH_2*COOH$$

Procedure[1]

To a solution of cyanoacetic-1-^{13}C acid (1.70 g, 20 mmol) in water (25 ml) is added hydrochloric acid (conc., 6 ml). Following hydrogenation at 50 psig over platinum oxide (0.25 g) (theoretical uptake after 16 hr), the solution is filtered and evaporated at reduced pressure to give a pale-yellow solid (3.0 g, Note 1). A solution of the hydrochloride in water (40 ml) is applied to IR-4B resin [25 ml, previously washed successively with hydrochloric acid (1%, 250 ml), sodium hydroxide (1%, 250 ml), and water (500 ml)], and eluted with water (250 ml). Removal of water at reduced pressure gives the product as a pale yellow solid (1.78 g, 100% yield).

Notes

1   The theoretical yield of β-alanine-1-^{13}C hydrochloride is 2.47 g; thus not all the water had been removed from the hygroscopic solid.

References

[1]J. L. Roberts and C. D. Poulter, "2',3',5'-Tri-O-benzoyl[4-^{13}C]uridine. An Efficient, Regio-specific Synthesis of the Pyrimidine Ring," *J. Org. Chem.*, **43**, 1547-1550 (1978).

DL-TYROSINE-3',5'-^{13}C$_2$

$$CH_3O-\underset{*}{\overset{*}{\bigcirc}}-CH_2Br \xrightarrow[\substack{2.\ H_2O \\ 3.\ HBr \\ 4.\ NH_4OH}]{\substack{1.\ AcNHC(Na)(COOEt)_2, \\ EtOH}}$$

$$HO-\underset{*}{\overset{*}{\bigcirc}}-CH_2CH(NH_2)COOH$$

Procedure[1]

1-(Bromomethyl)-4-methoxybenzene-3,5-^{13}C$_2$ (4.3 g, 21.2 mmol) is added to an ice-cold solution prepared from sodium (0.53 g, 23 mmol), ethanol (45 ml), and diethyl acetamidomalonate (4.98 g, 23 mmol), and the mixture is stirred for 5 hr. Water (120 ml) is added, and the precipitated solid is filtered and dried to give diethyl 2-acetamido-2-(4-methoxybenzyl-3,5-^{13}C$_2$)malonate, mp 96-97° (6.7 g, 93% yield). The ester (5.95 g, 17.6 mmol) is refluxed in hydrobromic acid (48%, 60 ml) for 4 hr, and the solution is filtered and evaporated to dryness at reduced pressure. The residue is dissolved in water (35 ml), the pH is adjusted to 7 with ammonium hydroxide (conc.), and the precipitate is filtered and dried under vacuum to give the product (3.0 g, 93% yield) (Note 1).

Notes

1 The product is resolved[2] through the N-trifluoroacetyl derivative and carbo-xypeptidase A PMSF to afford L-tyrosine-3',5',-^{13}C$_2$,[α]$^{24}_{546}$ −10.0° (c 2.0, 1 N HCl) (83% yield), and D-tyrosine-3',5'-^{13}C$_2$, [α]$^{24}_{546}$ + 10.1° (c 2.0, 1 N HCl) (63% yield).

Other Preparations

By the same procedure diethyl 2-acetamido-2-(4-methoxybenzyl)malonate-2-^{13}C and DL-tyrosine-2-^{13}C have been prepared.[2]

References

[1]V. Viswanatha and V. J. Hruby, "Synthesis of [3',5'-^{13}C$_2$] Tyrosine and Its Use in the Synthesis of Specifically Labeled Tyrosine Analogues of Oxytocin and Arginine-Vasopressin and Their 2-D-Tyrosine Diastereoisomers," *J. Org. Chem.*, **44**, 2892-2896 (1979).

[2]M. Blumenstein, V. J. Hruby, and D. M. Yamamoto, "Evidence from Hydrogen-1 and Carbon-13 Nuclear Magnetic Resonance Studies That the Dissociation Rate of Oxytocin from Bovine Neurophysin at Neutral pH Is Slow," *Biochemistry*, 17, 4971-4977 (1978).

3-(3,4-DIHYDROXYPHENYL)-2-METHYLALANINE-3-[13]C
(^{13}C-α-METHYLDOPA)

PhCH$_2$O—⟨phenyl⟩—*CH$_2$COCH$_3$
PhCH$_2$O—

$\xrightarrow{\text{KCN}}$ (NH$_4$)$_2$CO$_3$

PhCH$_2$O—⟨phenyl⟩—*CH$_2$—C(CH$_3$)—C=O
PhCH$_2$O—   HN   NH
                                    |
                                    C=O

$\xrightarrow[\text{2. HCl}]{\text{1. NaOH, diglyme}}$

HO—⟨phenyl⟩—*CH$_2$C(CH$_3$)COOH
HO—   NH$_2$·HCl

Procedure[1]

5-[3,4-Bis(benzyloxy)(benzyl-α-^{13}C)]-5-methylhydantoin A mixture of 1-[3,4-bis(benzyloxy)phenyl]-1-propanone-1-[13]C (7.5 g, 21.7 mmol), ethanol (150 ml), water (40 ml), potassium cyanide (1.8 g, 28 mmol), and ammonium carbonate (18.25 g, 190 mmol) is heated under reflux with stirring for 6.5 hr (Note 1). After standing overnight at room temperature, additional water (20 ml) is added, and most of the ethanol is evaporated. The white solid is filtered, washed with water, and dried to give the crude product, mp 182-184° (8.57 g, 95% yield) (Note 2), which is recrystallized from ethanol (97% recovery).

3-(3,4-Dihydroxyphenyl)-2-methylalanine-3-[13]C A mixture of the above-mentioned hydantoin (0.87 g, 2.1 mmol), water (4 ml), and diglyme (2 ml) containing sodium hydroxide (2.0 g, 50 mmol) is heated under reflux with stirring at 125° for 6 hr. The hot, heterogeneous reaction mixture is transferred to a 30-ml separatory funnel and allowed to cool to approx. 40°. The upper, yellow, organic layer is removed, and the white solid that forms on cooling is suspended in ether (20 ml), filtered, and washed with ether (10 ml). The solid is dissolved in water (30 ml) at 45°, and the solution is filtered. The pH is adjusted to 5.5 with acetic acid (5%), and the resulting precipitate is filtered, washed with cold water (10 ml), dispersed in hot ethanol (30 ml), cooled, filtered, and washed with cold ether and ethanol to give 3-[3,4-bis(benzyloxy)phenyl]-2-methylalanine-3-^{13}C, mp (dec.) 219-231° (0.56 g, 68% yield) (Note 2).

The solid is transferred to a 3-neck 50-ml flask, benzene (10 ml) and hydrochloric acid (conc., 10 ml) are added, and the heterogeneous mixture is vigorously stirred at room temperature under a nitrogen atmosphere for 18 hr. The solvents are removed and the hygroscopic white solid is dried under vacuum (0.35 g, 100% yield) (Note 2).

Notes

1 Solids that form in the condenser are washed down with water.
2 ^{1}H-NMR and mass spectral data are given.

References

[1]M. M. Ames and N. Castagnoli, Jr., "The Synthesis of ^{13}C-Enriched α-Methyldopa," *J. Labelled Compds.*, **10**, 195-205 (1974).

2-METHYLPROPENENITRILE-1-^{13}C or 2-METHYLPROPENENITRILE-2-^{13}C (^{13}C$_1$-METHACRYLONITRILE)

$$CH_3{}^\bullet COCH_3 + K^*CN \xrightarrow[H_2O]{H_2SO_4} (CH_3)_2{}^\bullet C(OH)^*CN$$

$$\xrightarrow[170°]{P_2O_5} CH_2={}^\bullet C(CH_3)^*CN$$

Procedure[1]

2-Hydroxy-2-methylpropionitrile-1-^{13}C or 2-Hydroxy-2-methylpropionitrile-2-^{13}C
To a 100-ml 4-neck flask fitted with a stirrer, thermometer, gas inlet, and dropping funnel and containing acetone (5.5 g, 95 mmol), water (10 ml), and potassium cyanide (5.9 g, 95 mmol) is added dropwise over 2 hr sulfuric acid (conc., 5.3 ml) in water (14.7 ml) with ice cooling and stirring. The temperature is kept below 15° during the addition. The reaction mixture is extracted with ether (4 × 20 ml), and the extract is dried over sodium sulfate, evaporated, and distilled at reduced pressure. The forerun, containing hydrogen cyanide, is dissolved in aqueous sodium hydroxide, and the main fraction, bp 82° at 23 mm Hg, contains the product (7 g, 75% yield).

2-Methylpropenenitrile-1-^{13}C or 2-Methylpropenenitrile-2-^{13}C Phosphorus pentoxide (20 g) and a spatula tip of hydroquinone are placed in a 200-ml 3-neck flask fitted with a magnetic stirrer and condenser and cooled in an ice bath. The ^{13}C-acetone cyanohydrin (7 g, 80 mmol) is added and stirred approx. 5 min. The mixture is heated quickly in a metal bath at 170°. Following distillation of some hydrogen cyanide and acetone, the main fraction, bp 75-100°, is obtained as the bath is heated to 250°. The major fraction obtained on redistillation (3.8 g) consists of acetone (10%) and the product (90%) (56% yield).

References

[1]B. Stuetzel, W. Ritter, and K-F. Elgert, "^{13}C-Enriched Monomers for NMR-Spectroscopy of Polymers," *Angew. Makromolec. Chem.*, **50**, 21-41 (1976).

2-(CYANO-^{13}C)PROPIONIC ACID

$$Na^*CN + CH_3CHBrCOONa \xrightarrow[\text{2. HCl}]{\text{1. } H_2O} CH_3CH(^*CN)COOH$$

Procedure[1]

Sodium 2-bromopropionate (28.0 g, 0.160 mol) (Note 1), sodium cyanide-^{13}C (7.99 g, 0.160 mol), and sodium hydroxide (0.48 g, 0.012 mol) in water (28 ml) are heated to 50° and stirred 2.5 hr. The mixture is cooled, acidified with hydrochloric acid (10%, 55 ml), and extracted with ether (3 × 75 ml). The extract is dried over magnesium sulfate, filtered, and evaporated to give the product as a yellow oil (13.4 g, 84% yield).

Notes

1 The salt is prepared by treating 2-bromopropionic acid (24.5 g, 0.160 mol) in acetonitrile (100 ml) with sodium carbonate (8.48 g, 0.080 mol); the precipitated solid is filtered, washed with ether, and dried.

References

[1]C. M. Redwine and T. W. Whaley, "Syntheses with Stable Isotopes: Thymine-2,6-^{13}C$_2$," *J. Labelled Compds.*, **16**, 315-320 (1979).

2-(CYANO-^{13}C)PROPIONIC-3,3,3-d_3 ACID

$$N^*CCH_2COOMe \xrightarrow[\text{NaH, THF}]{CD_3I} N^*CCH(CD_3)COOMe$$

$$\xrightarrow[\text{2. HCl}]{\text{1. NaOH, MeOH}} N^*CCH(CD_3)COOH$$

Procedure[1]

Methyl 2-(cyano-^{13}C)propionate-3,3,3-d_3 A slurry of sodium hydride (0.13 mol) in tetrahydrofuran (200 ml) is added over 20 min to an ice-cold, stirred mixture of methyl (cyano-^{13}C)acetate (Note 1) (20.0 g, 0.20 mol), iodomethane-d_3 (18.4 g, 0.13 mol), and tetrahydrofuran (200 ml). After stirring another 10 min the solvent is removed at reduced pressure, and the residue is treated with dichloromethane (200 ml), filtered, and evaporated to leave a yellow oil (21.7 g). Fractional distillation at 20 mm Hg through a spinning-band column gives the product, bp 73-75° (7.55 g), methyl (cyano-^{13}C)-2-(methyl-d_3)propionate-3,3,3-d_3, bp 63-69° (1.10 g), and unreacted starting material (8.31 g) in the pot residue (Note 2). Total yield of product is 8.47 g (65 % based on recovered starting material).

2-(Cyano-^{13}C)propionic-3,3,3-d_3 acid   A solution of the ester above (8.4 g, 72 mmol) in methanolic sodium hydroxide (6%, 100 ml) is kept at room temperature for 2 hr and evaporated at reduced pressure. The residue is dissolved in hydrochloric acid (20%, 90 ml), evaporated, and stirred with ether-acetone (1:1, 200 ml). The solution is dried over magnesium sulfate, filtered, and evaporated to give the product as a pale-yellow oil (7.16 g, 97% yield).

Notes

1   The starting material is prepared according to a published procedure.[2]
2   Additional starting material (0.97 g) and additional product are recovered by preparative gas chromatography of intermediate product-rich fractions.

References

[1]J. A. Lawson, J. I. DeGraw, and M. Anbar, "Synthesis of Hexalabeled Thymine and Thymidine," *J. Labelled Compds.*, **11**, 489-499 (1975).
[2]A. Murray and D. L. Williams, *Organic Syntheses with Isotopes*, Interscience, New York, 1958, p. 441.

MALONIC-1-^{13}C ACID

$$CH_3*COOH \xrightarrow[\begin{array}{l} \text{1. } Br_2, P, Ac_2O \\ \text{2. } Na_2CO_3, NaCN, H_2O \\ \text{3. } NaOH \\ \text{4. } HCl \end{array}]{} HOOCCH_2*COOH$$

Procedure[1]

The acetic-1-^{13}C acid obtained from sodium acetate-1-^{13}C (2.46 g, 30 mmol) (Note 1), acetic anhydride (0.05 ml, 0.5 mmol) (Note 2), and red phosphorus (0.01 g) in a 10-ml 2-neck flask fitted with a reflux condenser and calcium chloride tube are heated to 125-135° in an oil bath. Bromine (2.5 ml, 49 mmol) is carefully added through the condenser, and the mixture is refluxed for 3 hr. The temperature is lowered to 50°, dry carbon dioxide is passed through the mixture to remove excess bromine, and the residue is taken up in water (30 ml) and transferred to a 100-ml flask. The pH is adjusted to 8 with sodium carbonate, sodium cyanide (3 g, 61 mmol) in water (20 ml) is slowly added to the stirred solution, and the mixture is heated for 1 hr on the steam bath. After cooling slightly, sodium hydroxide (3 g, 75 mmole) is added and the mixture is again heated on the steam bath for 2 hr.

Steam is passed through the solution to remove ammonia, and the solution is evaporated to approx. 50 ml, cooled in an ice bath, treated with hydrochloric acid (conc., approx. 10 ml), and extracted with ether (3 × 50 ml). The combined extracts are dried over sodium sulfate, filtered, and evaporated at reduced pressure to give the product, which is sublimed at 90-100° at 1 mm Hg, mp 130-133° (1.32 g, 42% yield) (Note 2).

Notes

1 The acid is obtained by heating the salt with polyphosphoric acid (10 ml) at 170-180°.
2 The yield and isotopic enrichment must be corrected for the acetic anhydride used. Use of trifluoroacetic anhydride, or other procedures, would obviate the isotopic dilution.

References

[1]D. L. Fitzell, D. P. H. Hsieh, C. A. Reece, and J. N. Seiber, "Preparation of Acetic-1-^{13}C and Malonic-1-^{13}C Acids for Biosynthetic Studies," *J. Labelled Compds.*, 11, 135-139 (1975).

DIETHYL MALONATE-1-^{13}C

$$BrCH_2*COOH \xrightarrow[\text{2. EtOH, H}_2\text{SO}_4]{\text{1. NaOH, NaCN}} EtOOCCH_2*COOEt$$

Procedure[1]

The bromoacetic-1-^{13}C acid obtained from acetic-1-^{13}C acid (10.15 g, 169 mmol) is neutralized, under external cooling, with sodium hydroxide (33%), and an additional amount (0.5 ml) of the base is added. To the stirred solution at 0° is added sodium cyanide (9.3 g, 190 mmol) in water (25 ml). After 1.5 hr the solution is heated to 95°, and the water is evaporated at reduced pressure. Ethanol (50 ml, 850 mmol) and benzene (25 ml) are added, the solution is cooled to 0°, and sulfuric acid (46 ml) is added dropwise over 5 min. The stirred mixture is refluxed for 15 hr, water (25 ml) and benzene (35 ml) are added, and the aqueous layer is extracted with benzene (2 × 20 ml). The combined extracts are washed with water (15 ml), saturated sodium bicarbonate solution (15 ml), and water (15 ml). After drying over magnesium sulfate, the solution is distilled at 12 mm Hg to give the product, bp 83.5-85.5° (19.4 g, 71% yield).

References

[1]B. Bak and J. J. Led, "Preparation of [1-^{13}C], [3-^{13}C], [1-D] and [3-D] Enriched Cyclobutenes," *J. Labelled Compds.*, 4, 22-27 (1968).

SODIUM 2-(METHYL-^{13}C)BUTANOATE

$$*CH_3I + CH_3CH_2CH(COOEt)_2 \xrightarrow[\substack{3.\ H_2SO_4 \\ 4.\ NaOH}]{\substack{1.\ NaOEt,\ EtOH \\ 2.\ KOH}}$$

$$CH_3CH_2CH(*CH_3)COONa$$

Procedure[1]

To the sodium ethoxide solution prepared from sodium (0.49 g, 21.3 mmol) and ethanol (4 ml) (Note 1) at a temperature somewhat less than reflux is slowly added diethyl 2-ethylmalonate (3.23 g, 17.2 mmol) with vigorous stirring. After all the precipitated sodium ethoxide has dissolved stirring is continued another hour. While heating is interrupted methyl-^{13}C iodide (3.00 g, 21.0 mmol) is added very slowly (Note 2), and then the solution is refluxed for 3 hr. The reaction mixture is cooled, washed with water (10 ml), treated with potassium hydroxide (7.0 g, 0.125 mol) in water (10 ml), and refluxed for 12 hr with stirring. The mixture is cooled, treated with sulfuric acid (82.8 mmol) and refluxed for 3 hr. The solution is extracted with ether (6 X. equal volumes), and the extracts are dried over sodium sulfate and distilled to give 2-(methyl-^{13}C)butanoic acid, bp 120-150°. Titration to pH 9.0 with sodium hydroxide (1.0 N) and lyophilization gives the salt (1.65 g, 63% yield).

Notes

1   The reagent is prepared in an argon atmosphere under reflux at a temperature sufficient to melt the sodium; a yellow solution containing some white precipitate is obtained.

2   A milky white solution results.

3   Saponification of the dimethylmalonate required 24 hr.

4   Decarboxylation of this acid is more difficult than for other dialkylmalonates; heating in aqueous sulfuric acid gives incomplete reaction.

Other Preparations

2-Methylpropionic-1-^{13}C acid, butanoic-1-^{13}C acid, and 2-methylbutanoic-1-^{13}C acid have been prepared (as their sodium salts) in 74, 81, and 82% yields, respectively, by carbonation of the appropriate Grignard reagents with carbon-^{13}C dioxide.[1] Using a procedure similar to that given above, 2,2-di(methyl-^{13}C)malonic acid is prepared[1] using diethyl malonate (3.00 g, 18.7 mmol), sodium (0.96 g, 41.5 mmol), ethanol (10 ml), and methyl-^{13}C iodide (6.00 g, 42.0 mmol) (Note 3). The acid is recrystallized from ether-hexane, and the crystalline solid is decarboxylated by heating slowly to 195° (Note 4). The resulting 2-(methyl-^{13}C)propionic-3-^{13}C acid is distilled and converted to the sodium salt (1.03 g, 44% yield).

References

[1]B. H. Baretz, C. P. Lollo, and K. Tanaka, "Synthesis of Short Chain Carboxylic Acids Labelled with ^{13}C and ^{2}H at Various Positions," *J. Labelled Compds.*, **15**, 369-379 (1978).

METHIONINE-1-^{13}C

$$K*CN + CH_3SCH_2CH_2CHO \xrightarrow[\text{2. HCl}]{\text{1. NH}_3,\text{ EtOH}}$$

$$CH_3SCH_2CH_2CH(NH_2)*COOH$$

Procedure[1]

Ethanol (40 ml) is saturated with ammonia, and ammonium hydroxide (15 N, 35 ml), potassium cyanide-^{13}C (1.7 g, 26 mmol), and ammonium chloride (1.37 g, 26 mmol) are added. 3-(Methylthio)propionaldehyde (2.68 g, 26 mmol) is added to the stirred solution over a 5-min period, and stirring is continued at room temperature for 12 hr. The mixture is evaporated at reduced pressure until a light-yellow oil remains. Hydrochloric acid (12 N, 20 ml) is added, and the solution is heated at 100° for 4 hr. Water (100 ml) is added to the cooled reaction mixture and evaporated at reduced pressure. The process is repeated three times (Note 1). The residue is taken up in water (100 ml), treated with charcoal (1 g), heated to boiling, and filtered. The filtrate is neutralized with sodium hydroxide (2 N) and allowed to stand overnight at 10° to give a first crop of product (2.6 g). Evaporation of most of the water gives a second crop (0.7 g, total yield 84%), mp (dec.) 282° (Note 2).

Notes

1 Repeated evaporation of the water removes the hydrochloric acid.
2 Thin layer chromatography, infrared, and mass spectral data are given.

References

[1]B. D. Andresen, "Synthesis of *D,L*-1-^{13}C-Methionine," *J. Labelled Compds.*, **14**, 589-597 (1978).

$^{13}C_3$-VALINE HYDROCHLORIDE

$$*CH_3CN \xrightarrow[\text{EtOH}]{H_2NCH_2CH_2SH} \underset{*CH_3}{\overset{}{\text{S}\diagup\diagdown\text{N}}} \xrightarrow[\substack{\text{2. } *CH_3I \\ \text{3. BuLi} \\ \text{4. } *CH_3I}]{\text{1. BuLi, THF}}$$

$$\underset{*CH_3*CH*CH_3}{\overset{}{\text{S}\diagup\diagdown\text{N}}} \xrightarrow[\text{2. HgCl}_2, H_2O, MeCN]{\text{1. Al-Hg, }H_2O\text{-Et}_2O} (*CH_3)_2*CHCHO$$

$$\xrightarrow[\substack{\text{1. NaHSO}_3, \text{MeCN}, H_2O \\ \text{2. NaCN}, H_2O \\ \text{3. NH}_4OH \\ \text{4. HCl}}]{} (*CH_3)_2*CHCH(NH_2)COOH \cdot HCl$$

Procedure[1] (Note 1)

2-(Isopropyl-$^{13}C_3$)-2-thiazoline   A solution of acetonitrile-2-^{13}C (19.5 g, 0.47 mol) and aminoethanethiol (35.9 g, 0.47 mol) in ethanol (95 ml) is refluxed for 22.5 hr. After distillation of the alcohol the residue is treated with water (100 ml) and ether (100 ml), the layers are separated, and the aqueous phase is again extracted with ether (8 X 50 ml). The ether extract is dried over potassium carbonate and the ether is distilled. Distillation of the residue at reduced pressure affords 2-(methyl-^{13}C)-2-thiazoline, bp 43.5-44.5° at 17 mm Hg (33.6 g, 70% yield).

A solution of the thiazoline (52 g, 0.50 mol) in tetrahydrofuran is magnetically stirred in a nitrogen atmosphere, cooled to −78° (Dry Ice-isopropyl alcohol bath), treated with butyllithium (1.68 M in hexane, 300 ml, 0.50 mol) over a period of 40 min, and stirred an additional 1.5 hr. Methyl-^{13}C iodide (71 g, 0.50 mol) is added dropwise over 1.5 hr, stirring is continued for 1 hr, the temperature is allowed to increase to −10°, and stirring is continued for 2 hr. The reaction mixture is again cooled to −78°, and the above treatments with butyllithium and methyl-^{13}C iodide are repeated. The mixture is poured onto ice (1 kg), and the aqueous layer is adjusted to pH 2, extracted with hexane (200 ml), adjusted to pH 10 with sodium hydroxide, and extracted with hexane (4 X 200 ml). The hexane extract is dried over potassium carbonate and distilled to give the product as a colorless liquid, bp 62-65° at 15 mm Hg (56.4 g, 86% yield).

2-(Methyl-^{13}C)propanal-2,3-$^{13}C_2$ ($^{13}C_3$-**isobutyraldehyde**)   Aluminum foil (79 g, 2.9 mol), as 1.2-cm squares, in a 5-liter, round-bottom flask is treated with potassium hydroxide (5%, 1 liter) for 10 min, washed with water (4 liter), treated twice with mercuric chloride (0.5%, 900 ml, 17 mmol) for 3 min, and washed with water (5 liters), ethanol (95%, 2 X 1 liter), and ether (2 X 1.5 liters). A solution of the above $^{13}C_3$-thiazoline (56 g, 0.43 mol) in moist ether (2.9 liters) is added, and the mixture is refluxed for 2 hr, cooled, and filtered. The aluminum amalgam is washed with ether (2 X 500 ml), and the combined ether solutions are evaporated.

The residual oil is dried over potassium carbonate, filtered, and distilled to afford 2-(isopropyl-$^{13}C_3$)thiazolidine (47.6 g, 83% yield).

The thiazolidine (26.79 g, 0.200 mol) in acetonitrile (60 ml) is added over 40 min to a stirred solution of mercuric chloride (57.3 g, 0.211 mol), water (48 ml), and acetonitrile (192 ml) (Note 2), and stirring is continued an additional 2 hr. Sodium chloride (100 g) and water (190 ml) are added, a condenser is fitted for distillation, and the mixture is heated. The distillate, bp 64-70°, is dried over potassium carbonate, the acetonitrile solution of the aldehyde is added with stirring to a saturated solution of sodium bisulfite (20.8 g, 0.200 mol) in water (24 ml), and the mixture is stirred overnight. The bisulfite addition product is filtered, washed with acetonitrile, and dried (35.63 g, 99% yield).

2-Amino-3-(methyl-^{13}C)butanoic-3,4-$^{13}C_2$ acid hydrochloride To a solution of the $^{13}C_3$-isobutyraldehyde bisulfite addition compound (35.63 g, 0.200 mol) in water (79 ml) cooled in an ice bath is added a solution of potassium cyanide (26.85 g, 97%, 0.400 mol) in water (55 ml) over 10 min, and stirring is continued at room temperature for 2 hr. The solution is extracted with methylene chloride (5 × 50 ml), and the extract is dried over magnesium sulfate and evaporated at reduced pressure to give 2-hydroxy-3-(methyl-^{13}C)butanenitrile-3,4-$^{13}C_2$ (16.34 g, 80% yield). The cyanohydrin is added dropwise with stirring to ammonium hydroxide (conc., 51 ml), and the mixture is stirred for 6 hr at room temperature. The cloudy solution is extracted with methylene chloride (5 × 30 ml), and the extract is washed with hydrochloric acid (10%, 4 × 55 ml). The aqueous solution is evaporated at reduced pressure to give 2-amino-3-(methyl-^{13}C)butanenitrile-3,4-$^{13}C_2$ hydrochloride (21.34 g, 97% yield) (Note 3).

A solution of the aminonitrile hydrochloride (26.86 g, 0.196 mol) in hydrochloric acid (conc., 142 ml) is allowed to stand at room temperature for 25 hr and then heated to reflux for 7 hr (Note 4). The mixture is cooled, finally in ice, to give colorless crystals, which are collected, washed with cold concentrated hydrochloric acid, and dried to give the crude product (16.33 g) containing ammonium chloride. The filtrates are evaporated to give additional solid (23.27 g), which is combined with the first crop and extracted with hot ethanol and filtered to remove ammonium chloride (9.94 g, 95%). The product obtained from the ethanol solution is recrystallized from acetonitrile to give the pure amino acid hydrochloride (26.07 g, 85% yield).

Notes

1 The method is applicable to a wide variety of isotope isomers. Conversion of the aldehyde to the amino acid without isolation of intermediates has given erratic results. ^{13}C-NMR and IR data are given for the product and intermediates.

2 A white solid precipitates.

3 The aminonitrile may also be isolated by precipitation from the methylene chloride solution with hydrogen chloride.

4 The aminonitrile is hydrolyzed by conc. hydrochloric acid at room temperature

for 24 hr to the amide, which is further converted to the acid by refluxing; direct treatment of the nitrile with hydrochloric acid at reflux gives lower yields.

References

[1]T. W. Whaley, G. H. Daub, V. N. Kerr, T. A. Lyle, and E. S. Olson, "Syntheses with Stable Isotopes: DL-Valine-$^{13}C_3$," *J. Labelled Compounds*, **16**, 809-817 (1979).

4-(4-METHOXYPHENYL)BUTANOIC-1-^{13}C ACID

$$4\text{-MeOC}_6\text{H}_4(\text{CH}_2)_3\text{Br} + \text{K*CN} \xrightarrow[90\text{-}95°]{\text{DMSO}} 4\text{-MeOC}_6\text{H}_4(\text{CH}_2)_3\text{*CN}$$

$$\xrightarrow[\text{2. HCl}]{\substack{\text{1. NaOH, KOH,}\\ \text{HOCH}_2\text{CH}_2\text{OH}}} 4\text{-MeOC}_6\text{H}_4(\text{CH}_2)_3\text{*COOH}$$

Procedure[1]

4-(4-Methoxyphenyl)butanenitrile-1-^{13}C Potassium cyanide-^{13}C (1.0 g, 17.9 mmole) is added to a solution of 4-(3-bromopropyl)anisole (5.24 g, 22.9 mmol) in dimethyl sulfoxide (130 ml). The mixture is stirred for 3 hr at 90-95°, poured into water (1.5 liter), and extracted with hexane to give the product as an oil (2.8 g, 84% yield) (Note 1).

4-(4-Methoxyphenyl)butanoic-1-^{13}C acid The nitrile above (2.79 g, 15.1 mol) is added to a solution of sodium hydroxide (12.5 g) and potassium hydroxide (12.5 g) in ethylene glycol (185 ml) and water (125 ml), and the mixture is stirred for 6 hr at 115°. The chilled solution is washed with ether and acidified to give the product (2.46 g, 84% yield) (Note 2).

Notes

1 The crude product is 95% pure by GLC.
2 The crude product is sufficiently pure for use in subsequent reactions.

References

[1]G. H. Walker and D. E. Hathaway, "Synthesis of *N*-Phenyl-2-[1,4,5,8-^{14}C]naphthylamine, *N*-Phenyl-2-[8-^{13}C]naphthylamine, and *N*-[U-^{14}C]Phenyl-1-naphthylamine," *J. Labelled Compds.*, **12**, 199-206 (1976).

4-(1-PYRENYL)BUTANOIC-1-^{13}C ACID

$$\text{K*CN} + \text{Ar}(\text{CH}_2)_3\text{Br} \xrightarrow{\text{DMSO}} \text{Ar}(\text{CH}_2)_3\text{*CN}$$

$$\xrightarrow[\text{HCl}]{\text{HOAc}} \text{Ar}(\text{CH}_2)_3\text{*COOH}$$

Ar =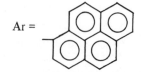

Procedure[1]

4-(1-Pyrenyl)butanenitrile-1-^{13}C To a well stirred mixture of potassium cyanide-^{13}C (500 mg, 7.56 mmol) and dimethyl sulfoxide (120 ml) at 70° is added 1-(3-bromopropyl)pyrene (2.45 g, 7.56 mol) (Note 1) over a 10-min period. The mixture is heated an additional 40 min, cooled, diluted with water (200 ml), and extracted with ether. The ether solution is dried over sodium sulfate, filtered, and evaporated at reduced pressure to give a yellow solid which is recrystallized from ethanol to give bright yellow crystals, mp 93-95° (1.41 g, 70% yield) (Note 2).

4-(1-Pyrenyl)butanoic-1-^{13}C acid The nitrile above (1.31 g, 4.86 mmol) is refluxed overnight with acetic acid (20 ml) and hydrochloric acid (conc., 10 ml). Water (100 ml) is added, and the precipitate is filtered, washed with water, dried under vacuum at 30-40°, and recrystallized from ethyl acetate to give mica-like crystals, mp 184-187° (1.1 g, 77% yield) (Note 2).

Notes

1 Directions are given for preparing the halide from 1-pyrenebutanoic acid.

2 TLC shows a single spot.

References

[1]W. P. Duncan, W. C. Perry, and J. F. Engel, "Labeled Metabolites of Polycyclic Aromatic Hydrocarbons IV. 7-Hydroxybenzo[a]pyrene-7-^{13}C," *J. Labelled Compds.*, **12**, 275-280 (1976).

SUCCINIC-1,4-^{13}C$_2$ ACID

$$2K^*CN + BrCH_2CH_2Br \xrightarrow[\text{(KI)}]{\text{EtOH}} N^*CCH_2CH_2^*CN$$

$$\xrightarrow[\substack{\text{2. KOH}}]{\substack{\text{1. H}_2\text{SO}_4 \\ \text{3. H}_2\text{SO}_4}} HOO^*CCH_2CH_2^*COOH$$

Procedure[1]

A mixture of potassium cyanide-^{13}C (19.3 g, 0.29 mol), a trace of potassium iodide, 1,2-dibromoethane (55 g, 0.29 mol), and ethanol (95 ml) is refluxed for 2.5 hr. Water (2 liter) is added, and the red mixture is extracted with ether (3 × 150 ml) (Note 1), acidified with sulfuric acid (5 N, 600 ml), and allowed to stand for 1 hr (Note 2). The solution (Note 3) is neutralized with potassium hydroxide, treated with an additional amount of the base (150 g), and refluxed for 20 hr. The mixture is acidified with sulfuric acid and continuously extracted with ether for 3 days to afford the product (13.6 g, 81% yield).

Notes

1 The ether extraction removes excess dibromoethane.
2 The acid treatment (caution: hydrogen cyanide odor) hydrolyzes any isonitriles to primary amines.
3 The intermediate succinonitrile-2,3-$^{13}C_2$ is not isolated.

Other Preparations

Succinic-2,3-$^{13}C_2$ acid has been prepared by a similar method from 1,2-dibromo-ethane-$^{13}C_2$ by way of succinonitrile-2,3-$^{13}C_2$.[2]

References

[1]B. Stuetzel, W. Ritter, and K-F. Elgert, "^{13}C-Enriched Monomers for NMR-Spectroscopy of Polymers," *Angew. Makromol. Chem.*, 50, 21-41 (1976).
[2]R. N. Renaud and L. C. Leitch, "Synthesis of 3,4,5,6-Tetradeuteriobenzene-1,2-$^{13}C_2$," *J. Labelled Compds.*, 9, 145-148 (1973).

DIETHYL SUCCINATE-2-^{13}C

$$Cl^*CH_2COOEt + CH_2(COOEt)_2 \quad \xrightarrow[\substack{2.\ NaOH, H_2O \\ 3.\ H_2SO_4}]{1.\ EtONa, EtOH}$$

$$HOOC^*CH_2CH_2COOH \xrightarrow[C_6H_6]{EtOH, HCl} EtOOC^*CH_2CH_2COOEt$$

Procedure[1]

Succinic-2-^{13}C acid To a refluxing solution of diethyl malonate (240 g, 1.5 mol) in a solution of sodium (35 g, 1.5 mol) in absolute ethanol (600 ml) is slowly added a solution of ethyl chloroacetate-2-^{13}C (47.5 g, 0.385 mol) in ethanol (100 ml) over a 1-hr period. Refluxing is continued another hour, sodium hydroxide (4 N, 750 ml) is added, and the mixture is refluxed for 12 hr more. The alcohol is evaporated, and the aqueous solution is transferred to a continuous extractor, acidified with sulfuric acid (18 N, 300 ml), and extracted with ether for 20 hr. The ether is distilled and the residue is heated at 150-160° for 2 hr, finally at reduced pressure. The crude product is dissolved in hot water (300 ml), treated with activated carbon, and filtered. Most of the water is evaporated to afford the crystalline product (28 g, 62% yield).

Diethyl succinate-2-^{13}C The above acid (28 g, 0.23 mol) is refluxed with hydrogen chloride-saturated ethanol (330 ml) and benzene (1.5 l) for 24 hr. Water formed during the reaction is removed in a trap. After distillation of the benzene and excess alcohol the residue is fractionated to give the product, bp 103° at 14 mm Hg (36.5 g, 94% yield).

Other Preparations

Diethyl succinate-1,4-^{13}C$_2$ has been prepared by esterification of the acid in the same manner.[1]

References

[1]B. Stuetzel, W. Ritter, and K-F. Elgert, "^{13}C-Enriched Monomers for NMR-Spectroscopy of Polymers," *Angew. Makromol. Chem.*, **50**, 21-41 (1976).

SUCCINIMIDE-^{15}N

*NH$_4$Cl + O=⟨⟩=O $\xrightarrow{200°}$ O=⟨⟩=O
            O
                              *N
                              H

Procedure[1]

Succinic anhydride (5.5 g, 55 mmol) and ammonium-^{15}N chloride (3 g, 55 mmol) are homogenized in a mortar and placed in a 50-ml flask that is coupled to a condenser by a short bend of tubing to avoid the possible reflux of water. The mixture is heated slowly in a Wood's metal bath at 2°/min to a final temperature of 200°. The temperature is maintained until the acidic vapors cease to distill (1-2 hr). The dark-colored mixture is dissolved in water, decolorized with activated carbon, neutralized with sodium carbonate, and left to crystallize. The crystals are dissolved in ethanol, treated with decolorizing carbon if required, crystallized and dried under vacuum (3.8 g, 70% yield).

References

[1]J. F. Arenas, V. Gomez, and R. Parellada, "Synthesis of Succinimide-^{15}N from ^{15}NH$_4$Cl," *An. Quim.*, **67**, 23-24 (1971).

MALEIC-1-^{13}C ANHYDRIDE

HOOCCH(NH$_2$)CH$_2$*COOH $\xrightarrow[\text{NaNO}_2, \text{H}_2\text{SO}_4]{\text{NaBr, HBr,}}$ HOOCHBrCH$_2$*COOH

$\xrightarrow[\text{MeCN}]{\text{(C}_6\text{H}_{11}\text{N)}_2\text{C}}$ O=⟨Br⟩=O $\xrightarrow{250\text{-}260°}$ O=⟨⟩=O

Procedure[1]

2-Bromosuccinic-4-^{13}C acid   To a stirred, ice-cooled solution of aspartic-4-^{13}C acid (1.01 g, 7.52 mmol) and sodium bromide (3.33 g, 32.3 mmol) in hydrobromic acid (2 N, 7 ml) is added over 1.75 hr pulverized sodium nitrite (0.93 g, 13 mmol). The mixture is stirred 15 min, treated with sulfuric acid (40%, 1.4 ml), stirred and cooled an additional 15 min, and extracted with ether (5 × 8 ml). The reddish-brown extracts are dried over magnesium sulfate, filtered, and evaporated to give the product, mp 164-165° (1.40 g, 94%).

Maleic-1-^{13}C anhydride   To the above acid (0.686 g, 3.48 mmol) in acetonitrile (8 ml) is added a solution of N,N-dicyclohexylcarbodiimide (0.729 g, 3.53 mmol) in acetonitrile (8 ml) (Note 1). After 1 hr the mixture is filtered, and the filtrate is evaporated at reduced pressure. The brownish oil is placed in a small tube that is in turn placed in a quartz tube (168 × 1.1 cm) positioned horizontally in the pyrolysis apparatus. A 5-cm-long oven surrounds the tube and is moved along by a pulley system at a speed of 11-12 mm/min over a distance of 160 cm while maintaining a temperature of 250° and helium or nitrogen flow of 35 ml/min. The liquid, which travels to the end of the tube with the furnace, solidifies on cooling, mp 49-52° (0.279 g, 82% yield) (Note 2).

Notes

1   Conversion to the anhydride before hydrogen bromide elimination is essential, because dehydrobrominations of bromosuccinic acid produce fumaric acid.

2   The product occasionally contains brown impurities which may be removed, with almost complete recovery of product, by repeating the pyrolysis.

Other Preparations

Maleic-2,3-^{13}C$_2$ has been prepared from bromosuccinic-2,3-^{13}C$_2$ acid by way of fumaric-2,3-^{13}C$_2$ acid.[2]

References

[1]J. M. Patterson, L. L. Braun, N. F. Haidar, J. C. Huang, and W. T. Smith, Jr., "A Novel Synthesis of Maleic Anhydride-1- and -2-^{13}C and Its Subsequent Conversion to Labeled Maleic Hydrazide," *J. Labelled Compds.*, **14**, 439-443 (1978).

[2]R. N. Renaud and L. C. Leitch, "Synthesis of 3,4,5,6-Tetradeuteriobenzene-1,2-^{13}C$_2$," *J. Labelled Compds.*, **9**, 145-148 (1973).

Pentanenitrile-^{15}N

$$KC*N + CH_3(CH_2)_3Cl \xrightarrow[120°]{DMSO, NaOAc} CH_3(CH_2)_3C*N$$

Procedure[1]

1-Chlorobutane is added dropwise over 40 min to a solution of potassium cyanide-^{15}N (1.09 g, 16.5 mmol) and sodium acetate (1.0 g) in dimethyl sulfoxide at 120°. The solution is refluxed for 18 hr, water is added, and the nitrile is extracted with three portions of ether. Hydrochloric acid is added (Note 1), and the aqueous solution is again extracted with ether. The ether is distilled, and the product is dried over phosphorus pentoxide (1.007 g, 73% yield).

Notes

1   The acid is added to hydrolyze isocyanate that may be present.

References

[1]J. B. Lambert, W. L. Oliver, and J. D. Roberts, "Nitrogen-15 Magnetic Resonance Spectroscopy. IV. The Degenerate Bimolecular Exchange of Protons in Ketimines," *J. Am. Chem. Soc.*, 87, 5085-5090 (1965).

5-AMINOLEVULINIC-5-^{13}C ACID

$$HCOCH_2CH_2COOEt \xrightarrow[\substack{2.\ K^*CN \\ 3.\ Ac_2O,\ C_5H_5N}]{1.\ NaHSO_3} AcOCH(^*CN)CH_2CH_2COOEt$$

Procedure[1]

Ethyl 4-acetoxy-4-(cyano-^{13}C)butyrate   Ethyl 4-oxobutyrate (0.5 g, 3.8 mmol) is added dropwise at 0° to a stirred solution of sodium metabisulfite (0.36 g, 1.9 mmol) in water (25 ml), and the solution is kept at 4° for 16 hr, cooled to 0°, treated dropwise with potassium cyanide-^{13}C (250 mg, 3.8 mmol) in water (1.3 ml), and stirred at 0° another hour. The solution is saturated with sodium chloride and extracted with benzene to afford the crude cyanohydrin, which is acetylated in pyridine (5 ml) with acetic anhydride (0.5 g, 5 mmol) at 20° for 20 hr. After addition of a few drops of water the pyridine is evaporated, and the residue in ether is washed with hydrochloric acid (2 N), sodium bicarbonate (saturated), dried, and evaporated to give the product, bp 105° at 0.3 mm Hg (0.7 g, 92% yield) (Note 1).

5-Hydroxy-2-piperidone-6-^{13}C The above ester (0.65 g, 3.3 mmol) in ethanol (20 ml) and hydrochloric acid (conc., 1.3 ml) is shaken under hydrogen at 20° and 1 atm with 5% palladium on carbon (0.65 g) for 1.5 hr (uptake 164 ml). The filtered solution is evaporated, and the residue is dried over phosphorus pentoxide, dissolved in ethanol (65 ml), treated with sodium hydride (50% dispersion, 169 mg, 7 mmol), stirred for 20 min at 20°, and neutralized (phenolphthalein) with acetic acid. The solution is filtered, evaporated, and chromatographed on alumina with methanol-chloroform (1:9) to give the product, which is recrystallized from ethanol ether, mp 144.5-145.5° (265 mg, 70% yield) (Note 1).

5-Aminolevulinic-5-^{13}C acid hydrochloride Pyridine (0.16 ml) and trifluoroacetic acid (0.08 ml) are added at 20° to a solution of the above piperidone (230 mg, 2.0 mmol) in dimethylsulfoxide (3.0 ml) containing N,N-dicyclohexylcarbodiimide (1.24 g, 6.0 mmol), and the mixture is kept at 20° for 1 day. Water (a few drops) and ethyl acetate (40 ml) are added, and the solution is filtered and evaporated to give a solution of piperidine-2,5-dione-6-^{13}C in dimethyl sulfoxide. Water (12 ml) and hydrochloric acid (conc., 24 ml) are added, and the solution is heated under reflux for 4 hr, evaporated to approx. 6 ml, diluted with water (60 ml), decolorized with charcoal, filtered, and evaporated (finally at 60° and 0.1 mm Hg). The residue in hydrochloric acid (0.05 N) is applied to a column of amberlite-IR-120(H$^+$) (25 ml), the column is washed with hydrochloric acid (0.05 N), and the product is eluted with hydrochloric acid (1 N, 150 ml) and recrystallized from ethanol-ether, mp 144-146° (297 mg, 89% yield) (Note 1).

Notes

1 IR, ^{1}H-NMR, and ^{13}C-NMR data are given.

References

[1]A. R. Battersby, E. Hunt, E. McDonald, and J. Moron, "Biosynthesis of Porphyrins and Related Macrocycles. Part II. Synthesis of δ-Amino[5-^{13}C]laevulinic Acid and [11-^{13}C]Porphobilinogen: Incorporation of the Latter into Protoporphyrin-IX," *J. Chem. Soc. Perkin I*, 1973, 2917-2922.

^{13}C-MEVALONOLACTONE

Method I

$$*CH_3COOH \quad \xrightarrow[\text{2. } PhCH_2OCH_2CH_2COCH_3]{\text{1. } BuLi/(i\text{-}Pr)_2NH, THF}$$

$$\underset{\underset{CH_3}{|}}{\overset{\overset{OH}{|}}{PhCH_2OCH_2CH_2C^*CH_2COOH}} \xrightarrow[\text{EtOH}]{H_2, Pd/C}$$

Procedure[1]

5-Benzyloxy-3-hydroxy-3-methylpentanoic-2-^{13}C acid Butyllithium (1.6 M, 40.6 ml, 65 mmol) in hexane is slowly added to a cooled solution of diisopropylamine (9.3 ml, 66.5 mmol) in tetrahydrofuran (25 ml). The mixture is stirred at 0-5° for 0.5 hr, and acetic-2-^{13}C acid (1.96 g, 32 mmol) (Note 1) in tetrahydrofuran (10 ml) is slowly added. The reaction mixture is stirred at 0-5° for 3.5 hr, 4-benzyloxy-2-butanone (5.34 g, 31 mmol) in tetrahydrofuran (15 ml) is added, and stirring is continued at 0-5° for 2 hr and then at room temperature for 18 hr. Water (20 ml) is added, the tetrahydrofuran is evaporated at reduced pressure, and ether is added. The mixture is extracted with sodium hydroxide (4%), and the aqueous phase is acidified with hydrochloric acid (18%) to pH 3 and extracted with ether. Evaporation at reduced pressure gives the product (4.01 g, 54% yield) (Note 2).

Mevalonolactone-2-^{13}C A solution of the above product (0.492 g, 2.06 mmol) in ethanol (95%, 50 ml) containing acetic acid (0.5 ml) (Note 3) and palladium black (0.10 g) is shaken in a hydrogenation apparatus for 2 hr at room temperature and 3.4 atm. The solution is filtered through Celite, additional palladium black (0.10 g) is added, and hydrogenation at 3.4 atm is continued for 8 hr at room temperature. The catalyst is removed by filtration, the ethanol is evaporated, and the product is distilled, bp 105° at 0.015 mm Hg (0.209 g, 91% yield) (Note 2).

Method II

$$(MeO)_2CHCH_2{^*}COCH_3 + Li{^*}CH_2COOMe \xrightarrow{\text{THF}}$$

$$\underset{\underset{CH_3}{|}}{\overset{\overset{OH}{|}}{(MeO)_2CHCH_2{^*}C^*CH_2COOMe}} \xrightarrow[\text{Et}_2\text{O}]{LiAlH_4} \underset{\underset{CH_3}{|}}{\overset{\overset{OH}{|}}{(MeO)_2CHCH_2{^*}C^*CH_2CH_2OH}}$$

$$\xrightarrow[\text{HOAc, H}_2\text{SO}_4]{\text{H}_2\text{O}_2}$$

Procedure[1]

Methyl 3-hydroxy-5,5-dimethoxy-3-methylpentanoate-2,3-^{13}C$_2$ To an ice-cold solution of acetic-2-^{13}C acid (1.00 g, 16.3 mmol) in tetrahydrofuran (5 ml) is slowly added diazomethane (2 M, 25 ml) in ether until a slight yellow color persists.

The solution is dried over magnesium sulfate (2 g), filtered, and quickly added by syringe to butyllithium (1.5 M, 11.0 ml, 16.5 mmol) in tetrahydrofuran (40 ml) containing hexamethyldisilazane (3.3 ml, 16.5 mmol) at $-78°$. The mixture is kept 50 min at $-78°$, treated with 4,4-dimethoxy-2-butanone-2-^{13}C (0.87 g, 6.60 mmol), after 10 min quenched with hydrochloric acid (3 N, 12 ml), and extracted with pentane (50 ml). The aqueous phase is extracted again with pentane (25 ml), and the total extract is dried over magnesium sulfate and evaporated to give the crude product as a yellow oil (1.28 g) (Note 4). Short-path distillation gives a colorless product (1.15 g, 83% yield) (Notes 1 and 5).

Mevalonolactone-3,4-^{13}C$_2$ A mixture of the above-mentioned ester (0.68 g, 3.3 mmol), lithium aluminum hydride (0.20 g, 5.2 mmol), and ether (40 ml) is stirred at room temperature for 1 hr. A solution of ethyl acetate (0.5 ml) in ether (2 ml) is added dropwise, followed by sodium bicarbonate solution (saturated, 0.5 ml). The precipitate is filtered and washed with ether, and the ether solution is dried and evaporated to give 5,5-dimethoxy-3-methyl-1,3-pentanediol-2,3-^{13}C$_2$ as a syrup (0.46 g, 78% yield). A mixture of the diol, acetic acid (1.5 ml), hydrogen peroxide (30%, 0.75 ml, 6.6 mmol), and sulfuric acid (1%, 2 ml) is heated at reflux for 1 hr, neutralized with potassium carbonate (70 mg), and evaporated at reduced pressure to give an oil (0.41 g), which is distilled to afford the product, bp 100° at 0.01 mm Hg (150 mg, 44% yield).

Notes

1 The acetic acid must be dry.
2 IR and NMR data are given.
3 Lactonization following hydrogenolysis is presumably catalyzed by the acetic acid.
4 Less than 5% of the starting ketone remains by GC analysis.
5 Yield is based on the ketone; the product is of sufficient purity for the next step.

Other Preparations

Mevalonic-2-^{13}C acid, dibenzylethylenediamine salt, has been prepared by sodium cyanoborohydride reduction of 3-hydroxy-5,5-dimethoxy-3-methylpentanoic-2-^{13}C acid obtained by barium hydroxide hydrolysis of the ethyl ester, which was synthesized by lithium amide condensation of ethyl acetate-2-^{13}C and 4,4-dimethoxy-2-butanone.[2] By a procedure similar to Method II mevalonolactone-4-^{13}C and mevalonolactone-3,4-^{13}C$_2$ have been prepared.[3]

References

[1]J. A. Lawson, W. T. Colwell, J. I. DeGraw, R. H. Peters, R. L. Dehn, and M. Tanabe, "The Synthesis of Singly and Doubly ^{13}C-Labeled Mevalonolactone," *Synthesis*, **1975**, 729-730.

[2]H. G. Floss, M. Tcheng-Lin, C. Chang, B. Naidoo, G. E. Blair, C. I. Abou-Chaar, and J. M. Cassady, "Biosynthesis of Ergot Alkaloids. Studies on the Mechanism of the Conversion of Chanoclavine-I into Tetracyclic Ergolines," *J. Am. Chem. Soc.*, 96, 1898-1909 (1974).

[3]D. E. Cane and R. H. Levin, "Application of Carbon-13 Magnetic Resonance to Isoprenoid Biosynthesis. II. Ovalicin and the Use of Doubly Labeled Malonate," *J. Am. Chem. Soc.*, 98, 1183-1188 (1976).

L-GLUTAMIC-^{15}N ACID

$$HOOCCOCH_2CH_2COOH \xrightarrow[\substack{\text{2. IR-120} \\ \text{3. IRA-400} \\ \text{4. HCl}}]{\substack{\text{1. *NH}_4Cl, \\ \text{Enzyme system}}}$$

$$L\text{-}HOOCCH(*NH_2)CH_2CH_2COOH \cdot HCl \longrightarrow L\text{-}HOOCCH(*NH_2)CH_2CH_2COOH$$

Glutamic decarboxylase

$$\xrightarrow[H_2O]{135°} \qquad \qquad H_2*N(CH_2)_3COOH$$

Procedure[1]

L-Glutamic-^{15}N acid hydrochloride A solution of potassium phosphate buffer (0.1 M, pH 7.6, 500 ml) and potassium 2-oxoglutarate (43 mmol) is adjusted to pH 7.6 with potassium hydroxide (5 M), and glucose 6-phosphate (sodium salt, 43 mmol), ammonium-^{15}N chloride (1.80 g, 33 mmol), adenosine 5-diphosphate (sodium salt, 0.5 mmol) (Note 1), and NADP (sodium salt, 70 μ) are added. The pH of the mixture is adjusted to 7.6 with either potassium hydroxide or hydrochloric acid, and the volume is increased to 800 ml with potassium phosphate buffer (0.1 M, pH 7.6). Portions of the mixture (200 ml) are placed in 200-ml volumetric flasks, and to each flask are added glucose 6-phosphate dehydrogenase (Sigma type 15, bakers' yeast, lyophilized, 100 IU enzyme units) and glutamic dehydrogenase (Sigma, bovine liver, type 2, 600 IU enzyme units). The four flasks are stoppered and incubated at 25° for 18 hr. The reaction mixtures are boiled for 2 min, filtered, and passed through a column of Amberlite IR-120(H⁺) resin (200 g) (Note 2). The product is eluted with potassium hydroxide (1 M) (Note 3) and passed through a column of Amberlite IRA-400(OH⁻) resin (100 g) (Note 2) from which it is eluted with hydrochloric acid (0.2 M). The solution of the product is concentrated at reduced pressure, saturated with hydrogen chloride, and kept at 5° overnight. The precipitated hydrochloride is filtered and washed with cold conc. hydrochloric acid and ethanol and dried (4.8 g, 80% yield) (Note 4).

L-Glutamic-^{15}N Acid   The free amino acid is obtained by elution of the IRA-400 column described above with acetic acid (0.2 M) instead of the hydrochloric acid, followed by concentration at reduced pressure and crystallization overnight at 5° (70-80% yield). Alternatively the hydrochloride may be neutralized with sodium hydroxide, or by exchange of chloride on an anion exchange column (OH$^-$ form).

L-Pyroglutamic-^{15}N acid   L-Glutamic acid (1 g) is placed in each of four screw-top Pyrex tubes containing water (15 ml). The tubes are sealed and heated at 135° in an oil bath for 3 hr. The combined solutions are evaporated to dryness at reduced pressure, and the residue is treated with ethanol (100 ml) at 65°. The solution is filtered (Note 5), concentrated to 25 ml at reduced pressure, treated with light petroleum (bp 40-60°, 75 ml), and kept at 5° overnight to give the product (2.18 g, 62% conversion) (Notes 4 and 6).

4-Aminobutanoic-^{15}N acid   L-Glutamic acid (4.9 g) is dissolved with heating and stirring in potassium acetate buffer (0.2 M, pH 5.0, 400 ml). The pH is adjusted to 4.7 with potassium hydroxide (5 M), glutamic decarboxylase (Sigma, type 2, 600 IU enzyme units) is added, and the solution is shaken at 37° for 4 hr. The solution is boiled, cooled, filtered, and applied to a column of Amberlite IRA-400(OH$^-$) anion exchange resin (130 g). The amino acid fraction that is eluted with hydrochloric acid (0.2 M) is treated with hydrochloric acid (5 M, 200 ml), evaporated to dryness at 60° at reduced pressure, and dried further over potassium hydroxide under vacuum. The solid is washed with ethanol-ether (1:1), filtered, dried over calcium chloride under vacuum, and dissolved in boiling ethanol (50 ml). The solution is kept at 5° overnight, and the crystals are filtered, washed with ethanol-ether, and dried (3.18 g, 68% yield).

Notes

1   The ADP is added to stabilize the glutamic dehydrogenase subsequently added.
2   The glutamic acid is adsorbed on the resin.
3   The ninhydrin reaction is used to follow the elution pattern.
4   The product is over 99% pure by GLC.
5   Unconverted glutamic acid (1.22 g, containing less than 3% pyroglutamic acid) remains on the filter and is recovered as the hydrochloride.
6   An improvement on this reaction, with nonisotopic materials, using hexamethyldisilazane in xylene has been reported (93% yield).[2]

References

[1]W. Greenaway and F. R. Whatley, "Enzymic Synthesis of L-Glutamic Acid-^{15}N and 4-Aminobutyric Acid-^{15}N and the Preparation of L-Pyroglutamic Acid-^{15}N," *J. Labelled Compds.*, **11**, 395-400 (1975).

[2]R. Pellegata, M. Pinza, and G. Pifferi, "An Improved Synthesis of γ-, δ-, and ε-Lactams," *Synthesis*, 1978, 614-616.

ADIPONITRILE-$^{15}N_2$

$$4*NH_3 + ClCO(CH_2)_4COCl \xrightarrow[H_2O]{CHCl_3} H_2*NCO(CH_2)_4CO*NH_2$$

$$\xrightarrow[MeCN]{POCl_3, \; AlCl_3} *NC(CH_2)_4C*N$$

Procedure[1]

Adipamide-$^{15}N_2$ A solution of ammonia-^{15}N (7 g, 0.47 mol) in water (18 ml) is cooled in an ice-salt bath and treated slowly with a solution of adipoyl chloride (12.5 g, 68 mmol) in chloroform (15 ml). The precipitate is filtered, washed with water, and recrystallized from water to give the product, mp 221-222° (6.4 g, 65% yield) (Note 1).

Adip[onitrile-$^{15}N_2$ A mixture of the amide above (6 g, 41 mmol), anhydrous aluminum chloride (0.25 g, 1.9 mmol), phosphorus oxychloride (7 g, 45 mmol), and acetonitrile (33 ml) is boiled and stirred 5 hr. The cooled mixture is filtered, the precipitate is washed with acetonitrile (2 X 15 ml), the solvent is distilled, and the residue is fractionated at reduced pressure to afford the product, bp 135-136° at 6 mm Hg (2.25 g, 50% yield).

Notes

1 The percent of theoretical yield given is based on adipoyl chloride; presumably, the large excess of ammonia-^{15}N and by-product ammonium-^{15}N chloride are recovered.

References

[1]O. G. Strukov, S. K. Smirnov, S. S. Dubov, and L. L. Danilana, "^{15}N-Substituted Amides and Nitriles VI. Synthesis and IR Spectra of ^{15}N-Substituted Adipamide and Adiponitrile," *Zh. Org. Khim.*, 7, 693-695 (1971) (English translation, pp. 703-705).

ADIPIC-$^{18}O_4$ ACID

$$4H_2*O + NC(CH_2)_4CN \xrightarrow{HBr} H*O*OC(CH_2)_4C*O*OH$$

Procedure[1]

Water-^{18}O (6.0 ml, 0.32 mol) in a 25-ml flask fitted with a condenser and drying tube and cooled in ice is saturated with hydrogen bromide. Adiponitrile (5.4 g, 50 mmol) is added in one portion with stirring. After the vigorous reaction has abated the mixture is heated on a steam bath 5 or 6 hr or overnight. Unreacted water-^{18}O is recovered on a vacuum line at $-78°$, and the residue of crude product is crushed with a glass rod, transferred to a Soxhlet extractor thimble, and extracted with ether for 6-8 hr. Evaporation of the ether gives the product, mp 152-153° (7.0 g, 93% yield).

References

[1]T.-S. Chen, J. C. Stephens, and L. C. Leitch, "Synthesis of 3-Deuteriocyclopentanone and Cyclopentanone-^{18}O," *J. Labelled Compds.*, 6, 174-178 (1970).

^{15}N$_1$-LYSINE

$$\text{phthalimide-*NK} + \text{PhCONH(CH}_2)_4\text{CHBrCOOEt}$$

$$\xrightarrow[\text{2. HCl, HOAc}]{\text{1. DMF}} \text{H}_2\text{N(CH}_2)_4\text{CH(*NH}_2\text{)COOH·HCl}$$

$$\text{phthalimide-*NK} + \text{Br(CH}_2)_4-\text{CH}\begin{matrix}\text{CO-NH}\\ | \\ \text{NH-CO}\end{matrix}$$

$$\xrightarrow[\substack{\text{2. Ba(OH)}_2\\ \text{3. HCl}}]{\text{1. DMF}} \text{H}_2\text{*N(CH}_2)_4\text{CH(NH}_2\text{)COOH·HCl}$$

Procedure[1]

Lysine-*N*^{2}-^{15}N   To ethyl 6-benzamido-2-bromohexanoate (8.25 g, 25 mmol) in dimethylformamide (50 ml) is added potassium phthalimide-^{15}N (3.7 g, 20 mmol), and the mixture is heated at 90° for 90 min with stirring. The mixture is diluted with chloroform (50 ml) and poured into water (120 ml). The chloroform is separated, the aqueous phase is extracted with chloroform (2 × 20 ml), and the combined extracts are washed with sodium hydroxide (0.1 N, 20 ml) and water (20 ml), dried over sodium sulfate, and evaporated at reduced pressure. The residue is hydrolyzed by heating at reflux with hydrochloric acid (conc., 40 ml) and acetic acid (20 ml) for 24 hr. The solution is cooled, filtered from the phthalic acid, and evaporated at reduced pressure. Water is added and again evaporated to remove excess hydrochloric acid. The residue is dissolved in water (50 ml), and the solution is filtered, adjusted to pH 5 with pyridine, and diluted with ethanol (4 volumes). After 24 hr at 4° the crystallized product is filtered, washed with ethanol, dried, and recrystallized from ethanol-water (8:1) (2.7 g, 73% yield) (Note 1).

Lysine-N^6-^{15}N 5-(4-Bromobutyl)hydantoin (9.4 g, 40 mmol), potassium phthalimide-^{15}N (6.2 g, 33 mmol), and dimethylformamide (75 ml) are heated at 90° for 90 min with stirring. The dimethylformamide is evaporated at reduced pressure, and the residue is washed with water (3 X 30 ml). Chloroform washings of the aqueous extracts are added to the reaction residue and evaporated at reduced pressure. The remaining oil is hydrolyzed in an autoclave with barium hydroxide (31 g of octahydrate) in water (170 ml) at 160° for 1 hr. After cooling the solution is treated with ammonium carbonate (12 g), filtered, and evaporated. The residue is refluxed with hydrochloric acid (conc., 100 ml) and acetic acid (50 ml) for 24 hr, and the product is isolated as in the preceding method (3.12 g, 51% yield) (Note 1).

Notes

1 Purity is established by electrophoresis and two-dimensional chromatography. Resolution of the racemic mixture is made through enzymatic hydrolysis with acylase I enzyme of the bis(N-chloroacetyl) derivative,[2] which is obtained by acylation of the DL-lysine with 4-nitrophenyl chloroacetate.[1]

References

[1] J. Mizon and Ch. Mizon, "Synthesis of DL-Lysine Selectively Labeled with Nitrogen-15 at the α- or ε-Positions and their Resolution," *J. Labelled Compds.*, **10**, 229-237 (1974).
[2] J. P. Greenstein and M. Winitz, *Chemistry of the Amino Acids*, Vol. 3, Wiley, New York, 1961, pp. 2097-2124.

DODECANEDIOIC-1,12-^{13}C$_2$ ACID

$$2Na^*CN + Br(CH_2)_{10}Br \xrightarrow[\text{2. HCl}]{\text{1. H}_2\text{O, EtOH}} HOO^*C(CH_2)_{10}^*COOH$$

Procedure[1]

1,10-Dibromodecane (3.62 g, 12 mmol) is added over a period of 10 min to a solution of sodium cyanide-^{13}C (1.0 g, 20 mmol) in water (1.5 ml) and ethanol (7.0 ml). The mixture is refluxed for 5 hr (Note 1), treated with hydrochloric acid (conc., 5.0 ml), refluxed for a further 12 hr, cooled, and extracted with ether. The extract is dried over sodium sulfate and evaporated to give the product, which is recrystallized from water, mp 127-128° (3.3 g, 86% yield).

Notes

1 The intermediate dodecanedinitrile-1,12-^{13}C$_2$ is not isolated.

References

[1] G. J. Shaw and G. W. A. Milne, "Synthesis and Properties of Dodecane-1,12-^{13}C$_2$ and Dodecane-1,1,1,12,12,12-^{2}H$_6$," *J. Labelled Compds.*, **12**, 557-563 (1976).

^{13}C-BENZOIC ACID

Method I

$$*CO_2 + PhMgBr \xrightarrow[\text{2. HCl}]{\text{1. } Et_2O, -20°}$$

$$Ph*COOH \xrightarrow[C_6H_6, DMF]{SOCl_2} Ph*COCl$$

Procedure[1]

Benzoic-^{13}C acid Carbon-^{13}C dioxide (4.5 liter, 0.2 mol) is added to a solution of phenylmagnesium bromide at $-20°$, which is prepared from bromobenzene (31 g, 0.2 mol) and magnesium (5.3 g, 0.22 mol) in ether (800 ml). After 10 min carbon dioxide is no longer absorbed and a dark gum precipitates. After acidification of the reaction mixture with hydrochloric acid (6 N), the ether solution is separated, washed with water, and extracted with sodium hydroxide (5%). The sodium hydroxide layer is acidified, and the precipitated product is collected and dried (17.7 g). The filtrate is washed with methylene chloride, and the extract is dried over sodium sulfate and evaporated to give additional product (2.8 g, 85% total yield).

Benzoyl-^{13}C chloride A solution of benzoic-^{13}C acid (72.5 g, 0.59 mol), thionyl chloride (47 ml, 0.65 mol), and dimethylformamide (0.3 ml, 4 mmol) in benzene (350 ml) is refluxed for 4 hr. The solvent is removed and the residue is distilled at reduced pressure to give the product, bp 106° at 40 mm Hg (64.9 g). The pot residue is refluxed with thionyl chloride (5 ml) and dimethylformamide (0.2 ml) in benzene (50 ml) for 18 hr. Removal of the solvent and distillation afford additional product (8.5 g, 89% total yield).

Method II

Procedure

Benzoic-^{13}C acid[2] Toluene-α-^{13}C (1.68 g, 18.0 mmol) and water (84 ml) are refluxed with magnetic stirring while potassium permanganate (6.80 g, 43.0 mmol) is slowly added through the reflux condenser over 0.5 hr. After refluxing overnight the mixture is cooled, filtered, evaporated to 10 ml, and acidified. The white crystals are filtered, washed twice with cold water, and dried; mp 123-124° (1.11 g). A second crop is isolated from the filtrate (0.22 g, 60% total yield).

Benzoic-1-^{13}C Acid[3]   A mixture of toluene-1-^{13}C (3.8 g, 45 mmol) and potassium permanganate (16 g, 0.10 mol) in water (150 ml) is refluxed for 6 hr, cooled, filtered, and acidified with sulfuric acid (9 *M*). After cooling in an ice bath the precipitated benzoic acid is filtered, and the filtrate is extracted with benzene for recovery of additional product (total yield 3.8 g, 74%).

Methyl Benzoate-1-^{13}C[3]   To benzoic-1-^{13}C acid (3.8 g, 31 mmol) in methanol-water (10:1, 50 ml) at room temperature is added an ether solution of diazomethane until a weak yellow color persists. After the ether is evaporated at reduced pressure the residue is dried over calcium hydride and distilled, bp 83-85° at 15 mm Hg (3.5 g, 82% yield).

Other Preparations

Benzoic-1,3,5-^{13}C$_3$ has been prepared by permanganate oxidation of toluene-1,3,5-^{13}C$_3$ in 83% yield.[4] Benzoyl-^{13}C chloride is prepared on a 1-g scale in 80% yield by refluxing benzoic-^{13}C acid in thionyl chloride for 2 hr.[2] Benzoic-^{13}C acid has been obtained by a procedure similar to Method I.[5]

References

[1]J. A. Kepler, J. W. Lytle, and G. F. Taylor, "Synthesis of 5,5-Diphenylhydantoin-2,4,5-^{13}C$_3$," *J. Labelled Compds.*, **10**, 683-687 (1974).

[2]G. A. Braden and U. Hollstein, "Synthesis of 1-Phenyl-1-^{13}C-ethene-1-^{13}C (*trans*-Stilbene) and Derivatives," *J. Labelled Compds.*, **12**, 507-516 (1976).

[3]B. Stuetzel, W. Ritter, and K-F. Elgert, "^{13}C-Enriched Monomers for NMR-Spectroscopy of Polymers," *Angew. Makromol. Chem.*, **50**, 21-41 (1976).

[4]A. Heusler and T. Gaumann, "Synthesis of Aromatic and Alicyclic Six-Membered Rings, Labelled with ^{13}C in Positions 1,3,5," *J. Labelled Compds.*, **12**, 541-544 (1976).

[5]M. Pomerantz and R. Fink, "Synthesis of Diethyl 1-[(^{13}C)Methyl]-3-phenyl(1,3-^{13}C$_2$)bicyclo-[1.1.0]butane-exo,exo-2,4-dicarboxylate," *J. Labelled Compds.*, **16**, 275-286 (1979).

3,4-(METHYLENEDIOXY)BENZOIC-^{13}C ACID

Procedure[1]

A solution of butyllithium in hexane (15%, 0.96 g, 17 mmol) is added dropwise over 30 min to a solution of 1-bromo-3,4-(methylenedioxy)benzene (3.24 g, 16.2 mmol) in ether (25 ml) at −30 to −35°. Carbon-^{13}C dioxide, derived from barium carbonate-^{13}C (1.98 g, 10.0 mmol) and perchloric acid (60%, 10 ml), is introduced over 15 min into the stirred reaction mixture at −30°. Unreacted carbon dioxide is condensed with liquid nitrogen and reintroduced into the mixture at −20-5°. The reaction mixture is acidified with sulfuric acid (6 N) to pH 1 and filtered to remove the crystals of product that form. The filtrate is extracted with ether, and the ether is extracted with sodium bicarbonate solution (5%), which is then acidified with sulfuric acid (6 N) to give additional product. The combined product is washed with water and dried over phosphorus pentoxide in vacuum at 80°; mp 229° (1.2 g, 72% yield).

Other Preparations

By a similar procedure 3,4-bis(benzyloxy)benzoic-^{13}C acid has been prepared on a 5-mmol scale in average yield of 67%.[2]

References

[1]T. Nagasaki, Y. Katsuyama, and H. Minato, "Synthesis of Radioactive and Stable Isotope-Labelled 1-Ethyl-6,7-methylenedioxy-4(1H)-oxocinnoline-3-carboxylic Acids (Cinoxacin)," *J. Labelled Compds.*, **12**, 409-427 (1976).

[2]M. W. Ames and N. Castagnoli, Jr., "The Synthesis of ^{13}C-Enriched α-Methyldopa," *J. Labelled Compds.*, **10**, 195-205 (1974).

BENZAMIDE-^{15}N

$$*NH_4NO_3 + PhCOCl \xrightarrow[\text{NaOH}]{C_6H_6,\,H_2O} PhCO*NH_2$$

Procedure[1] (Note 1)

Ammonium-^{15}N nitrate (1.0 g, 12.3 mmol) in water (6 ml) at 6° is treated with benzene (0.5 ml), sodium hydroxide (1.14 g, 29 mmol) in water (5 ml) (Note 2), and benzoyl chloride (1.44 ml, 12.4 mmol) in benzene (50 ml). The mixture (Note 3) is agitated vigorously by magnetic stirring at room temperature for 2 hr, cooled, filtered, and extracted with chloroform (5 × 10 ml). The extract is dried over sodium sulfate and evaporated to afford the crystalline product, mp 127° (1.495 g, 99% yield).

Notes

1   This procedure, in which the ammonia-^{15}N is generated within the reaction mix-
    ture, also obviates formation and recovery of by-product ammonium-^{15}N salts.
2   The sodium hydroxide solution at 6° is added slowly, with the tip of the pipet
    kept under the benzene layer.
3   The reaction flask is stoppered.

References

[1]U. Hornemann, "Synthesis of 2-Amino-2-deoxy-D-glucose-^{15}N and of 2-Amino-2-deoxy-L-glucose-2-^{14}C," *Carbohydr. Res.*, **28**, 171-174 (1973).

BENZONITRILE-^{15}N

$$2\text{*NH}_3 + \text{PhCOCl} \xrightarrow{\text{Et}_2\text{O}} \text{PhCO*NH}_2 + \text{*NH}_4\text{Cl}$$

$$\text{PhCO*NH}_2 \xrightarrow[180°]{\text{AlCl}_3 \cdot \text{NaCl}} \text{PhC*N}$$

Procedure[1]

Benzamide-^{15}N Ammonia-^{15}N, generated from ammonium-^{15}N chloride (3.00 g, 55.0 mmol) (Note 1), is introduced over 1.5 hr into a solution of benzoyl chloride (4.38 g, 31.2 mmol) in ether (120 ml) cooled in a Dry Ice bath (Note 2). The mixture, after standing at room temperature for 12 hr, is filtered, the solid is washed with ethanol (5 × 18 ml), and the filtrate is concentrated to incipient crystallization at reduced pressure. The solution is treated with benzene (75 ml) and filtered at its boiling point. The residue is washed with hot benzene (3 × 35 ml). Two crops of product (1.582 g) are obtained from the solution, and acetone washing of the original solid affords additional material, mp 122.8-123.2° (1.428 g, 90% total yield) (Note 3).

Benzonitrile-^{15}N Benzamide-^{15}N (1.582 g, 13.0 mmol) and sodium aluminum chloride (2.50 g, 13.0 mmol) are mixed carefully in a 50-ml round-bottom flask equipped with a distillation head. The mixture is heated in an oil bath at 180° until effervescence ceases, and heating is continued with a flame until distillation of the product is finished (0.995 g, 74% yield).

Notes

1 The ammonium chloride, in concentrated aqueous solution, is added dropwise to a refluxing solution of sodium hydroxide (5.7 g) in water (14 ml). The ammonia is swept with nitrogen from the reaction flask through the reflux condenser and dried by passage through two potassium hydroxide drying tubes.

2 Absorption of ammonia is complete, as shown by a hydrochloric acid trap at the exit of the reaction flask.

3 The ammonium chloride (1.58 g) is recovered quantitatively.

Other Preparations

By a similar procedure 20 substituted benzamides-^{15}N have been prepared; experimental details are given for 3-methoxybenzamide-^{15}N.[2] Benzamide-^{15}N has also been prepared by the general procedure.[2,3]

References

[1] J. B. Lambert, W. L. Oliver, and J. D. Roberts, "Nitrogen-15 Magnetic Resonance Spectroscopy. IV. The Degenerate Bimolecular Exchange of Protons in Ketimines," *J. Am. Chem. Soc.*, 87, 5085-5090 (1965).

[2] T. Axenrod, P. S. Pregosin, M. J. Wieder, E. D. Becker, R. B. Bradley, and G. W. A. Milne,

"Nitrogen-15 Nuclear Magnetic Resonance Spectroscopy. Substituent Effects on ^{15}N–H Coupling Constants and Nitrogen Chemical Shifts in Aniline Derivatives," *J. Am. Chem. Soc.*, 93, 6536-6541 (1971).

[3]J. Volford and D. Banfi, "Synthesis and ^{15}N-Labeling of 2-Thiazolines and Analogous Thiazines," *J. Labelled Compds.*, 11, 419-426 (1975).

BENZOIC-$^{18}O_2$ ACID

$$PhCCl_3 + 2H_2{*}O \longrightarrow PhC{*}O{*}OH + 3HCl$$

Procedure[1]

A mixture of benzotrichloride (25 g, 0.13 mol) and water-^{18}O (10 g, 0.55 mol) is refluxed for 15 hr with protection from atmospheric moisture. The warm crystalline mass is evaporated at reduced pressure to expel hydrogen chloride and excess water. The residue is dissolved in sodium hydroxide (8 g, 0.2 mol) in water (50 ml), and the solution is concentrated. The precipitated sodium benzoate-$^{18}O_2$ is recrystallized from water, dried 2 hr at 100°, powdered, and again dried at 100° to give a first crop (approx. 15 g). A second crop (9 g) obtained from the filtrate is acidified with hydrochloric acid to afford benzoic-$^{18}O_2$ acid (6 g).

References

[1]M. Kobayashi and R. Kiritani, "Organic Sulfur Compound. X. The Mechanism of the Carboxylic Ester Formation from Carboxylic Sulfurous Anhydrides," *Bull. Chem. Soc. Jap.*, 39, 1782-1784 (1966).

3-BROMOBENZOYL-^{18}O CHLORIDE

Procedure[1]

Chlorine is bubbled through a mixture of 3-bromobenzaldehyde-^{18}O (15.2 g, 82 mmol) at room temperature for 85 min. The solvent is removed at reduced pressure, and the residual liquid is distilled to give the product (13 g, 73% yield) (Note 1).

Notes

1   The acid chloride is converted by reaction with sodium peroxide to bis(3-bromobenzoyl-^{18}O) peroxide (8.7 g, 79% yield).

References

[1]M. Kobayashi, H. Minato, and Y. Ogi, "Scrambling of Oxygen in the Diester Produced from Labeled Diacyl Peroxide-Olefin Reaction," *Bull. Chem. Soc. Jap.*, **43**, 1158-1160 (1970).

N-(BENZOYL-^{18}O)-O-tert-BUTYLHYDROXYLAMINE
[N-(1,1-Dimethylethoxy)benzamide-^{18}O]

$$PhC^*OCl + (CH_3)_3CONH_2 \cdot HCl \xrightarrow[\text{EtOH}]{\text{Imidazole}} PhC^*ONHOC(CH_3)_3$$

Procedure[1]

A chilled solution of benzoyl-^{18}O chloride (11.21 g, 80 mmol) in ethanol (10 ml) is added dropwise to a stirred solution of imidazole (10.9 g, 0.16 mol), *tert*-butyl-hydroxylamine hydrochloride (10 g, 80 mmol), and ethanol (60 ml) at 0° (Note 1). Addition is complete after 30 min, and the mixture is stirred at 0° for 12 hr. The solvent is evaporated at reduced pressure, and the residue is dissolved in chloroform (250 ml). The solution is washed with water (2 × 50 ml), hydrochloric acid (3 *M*, 50 ml), and water (50 ml), dried over magnesium sulfate, filtered, and evaporated at reduced pressure. Recrystallization of the residue (12.5 g, 81% yield) from a mixture of Skellysolve B and 95% ethanol (85:15, 100 ml) gives the product, mp 134-135°.

Notes

1   The apparatus is protected from atmospheric moisture with potassium hydroxide.

References

[1]T. Koenig, M. Deinzer, and J. A. Hoobler, "Thermal Decomposition of N-Nitrosohydroxyla-mines. III. N-Benzoyl-N-nitroso-O-*tert*-butylhydroxylamine," *J. Am. Chem. Soc.*, **93**, 938-944 (1971).

PHTHALIC-$^{13}C_1$ ACID

Procedure[1]

o-Toluic-^{13}C acid   The carbon-^{13}C dioxide, generated from barium carbonate-[13]C (7.0 g, 35 mmol) and sulfuric acid (conc., 75 ml), is condensed in a solution of *o*-tolylmagnesium bromide (106 mmol) in tetrahydrofuran (72 ml) cooled in liquid nitrogen. The reaction mixture is allowed to warm to room temperature, stirred for 30 min, and poured into dilute sulfuric acid. The acid solution is extracted three times with ether, the combined organic solutions are extracted with sodium hydroxide (2 N), and the ether-washed aqueous layers are acidified. Extraction with ether provides the product (4.8 g, 99% yield).

Phthalic-[13]C$_1$ acid   A solution of *o*-toluic-^{13}C acid (9.9 g, 71 mmol) in sodium hydroxide (1 N, 400 ml) is treated with potassium permanganate (35 g, 0.22 mol) and stirred and heated for 4 hr. After cooling, sodium sulfite (saturated solution) is added to discharge the purple color, and the precipitated manganese dioxide is removed by filtration. The filtrate is evaporated at 80° at reduced pressure until precipitation starts. The minimum amount of water is added to redissolve the precipitate at 80°, and the solution is poured into hydrochloric acid (150 ml). After standing overnight at 0°, the product is filtered, washed, and dried at 60° in vacuum; mp 205° (9.4 g, 78% yield).

Other Preparations

Using procedures previously developed and involving some 13 steps, the phthalic-[13]C$_1$ acid is an intermediate in preparation of the title compounds of the reference.[1] The single label is distributed among positions 5, 10, and 11 in the tricyclic ring; a second carbon-13 is introduced in the terminal side chain using sodium cyanide-[13]C.

References

[1]I. Midgley, R. W. Pryor, and D. R. Hawkins, "Synthesis of [[13]C$_2$]-Amitriptyline, Nortriptyline and Desmethylnortriptyline," *J. Labelled Compds.*, **15**, 511-521 (1978).

N-(DIMETHYLPHENYLSILYL)PHTHALIMIDE-[15]N

Procedure[1]

Potassium phthalimide-[15]N (0.401 g, 2.16 mmol) in tetrahydrofuran (50 ml) is treated slowly with stirring under a nitrogen atmosphere with chlorodimethylphenylsilane (0.375 g, 2.20 mmol). The mixture is heated to 60° for 15 min, cooled to 0°, filtered from the potassium chloride, and evaporated at reduced pressure. The product is extracted with benzene and recrystallized from benzene (0.358 g, 89% yield).

Other Preparations

Trimethylsilylphthalimide-^{15}N is prepared by the same procedure.[1]

References

[1]A. F. Janzen and E. A. Kramer, "Cyclic Imides. Preparation of Silylphthalimides and Characterization by Infrared and Mass Spectrometry," *Can. J. Chem.*, 49, 1011-1018 (1971).

4,4′-BIPHENYLDICARBOXAMIDE-^{15}N$_2$

Procedure[1]

A mixture of 4,4′-biphenyldicarbonyl dichloride (23.3 g, 83.5 mmol) (Note 1) and ammonium-^{15}N sulfate (12.3 g, 91.8 mmol) in methylene chloride (800 ml) is treated with triethylamine (38.1 g, 377 mmol) and stirred at room temperature in a stoppered flask for 72 hr (Note 2). The cream-colored mixture is poured into water (300 ml), and the methylene chloride is evaporated with a stream of air and gentle warming. The solid is filtered, washed with water, methanol, and ether, and dried to afford the product as a tan powder, mp > 300° (16.0 g, 79% yield).

Notes

1   The acid chloride is prepared from the acid using the thionyl chloride-dimethylformamide method.
2   This procedure, in which generation of ammonia-^{15}N is conveniently carried out within the reaction mixture, also eliminates the formation of by-product ammonium-^{15}N chloride.

References

[1]T. W. Whaley, "Syntheses with Stable Isotopes: Benzidine-^{15}N$_2$," *J. Labelled Compds.*, 14, 242-248 (1978).

CHENODEOXYCHOLIC-24-^{13}C ACID

1. Na*CN, DMSO
2. NaOH, EtOH
3. H$^+$

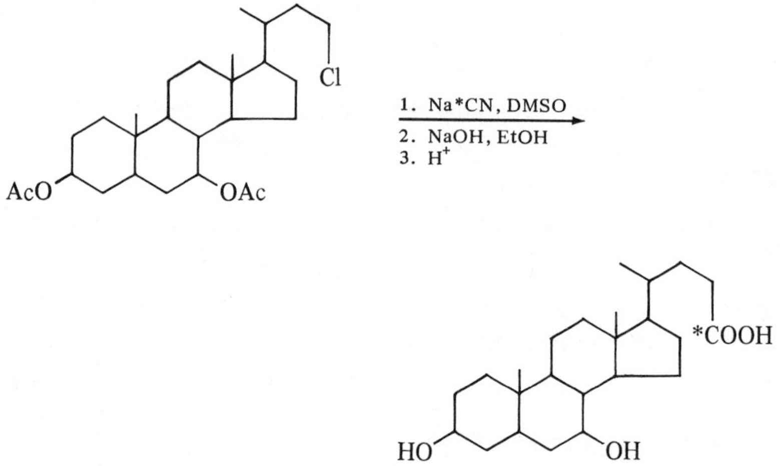

Procedure[1]

3α,7α-Diacetoxy-23-chloro-24-norcholane (2.82 g, 6.05 mmol) (Note 1) and sodium cyanide-^{13}C (0.260 g, 5.20 mmol) in dimethylsulfoxide are heated at 100° for 3 hr. The mixture is poured into water and extracted with ether to afford the crude nitrile as a white solid (2.76 g) (Note 2). The material is refluxed for 2 days in 90% ethanol (50 ml) containing sodium hydroxide (5.5 g). The mixture is poured into water, washed with ether, acidified, and extracted with ether. The extract is dried over magnesium sulfate, filtered through charcoal and Celite, and evaporated to afford a pale-yellow oil, which is recrystallized from ethyl acetate-hexane (1.48 g, 73% yield). A second recrystallization gives the pure product, mp 119-120° (1.21 g, 59% yield).

Notes

1   Directions are given for preparing the chloride from chenodeoxycholic acid.
2   The crude nitrile contains some chloride by TLC.

Other Preparations

Cholic-24-^{13}C acid, deoxycholic-24-^{13}C acid, and lithocholic-24-^{13}C acid have been prepared by way of the corresponding nitriles in the same manner.[1]

References

[1]D. L. Hachey, P. A. Szcepanik, O. W. Berngruber, and P. D. Klein, "Syntheses with Stable Isotopes: Synthesis of Deuterium and ^{13}C Labeled Bile Acids," *J. Labelled Compds.*, 9, 703-719 (1973).

3

Aldehydes and Ketones

Most reactions that have been used to produce carbon-13 labeled carbonyl compounds proceed from a carboxylic acid. Aldehydes can be prepared directly by certain metal hydride or catalytic reductions, or indirectly (but often more conveniently) through overreduction of an acid derivative to the alcohol followed by reoxidation. Oxidation of a secondary alcohol, from a Grignard reagent and an aldehyde (or other source), affords a ketone. Ketones are also obtained by a wide variety of acylation and condensation reactions, with the starting materials again being carboxylic acid derivatives. Oxygen-18 carbonyl compounds have been prepared by exchange or by introduction of the label through hydrolysis of a suitable derivative, such as an anil.

Several recently developed reactions that are likely to be applied to labeling of carbonyl compounds include the reaction of acid chlorides with Grignard reagents[1] and with organotin compounds using palladium catalysis,[2] the preparation of aldehydes by formylation of Grignard reagents[3] and by oxidation of trialkoxyboroxines obtained by reduction of carboxylic acids with borane-dimethyl sulfide,[4] the oxidation of trialkyl borates,[5] the further improvements in selective oxidations,[6] and the direct introduction of carbon monoxide into aldehydes through carbonylation of organoboranes[7] and with hybrid-phase catalysts.[8] Perhaps even the first two (and most difficult) members of the series can become more accessible.

References

[1] F. Sata, M. Inoue, K. Oguro, and M. Sato, "Preparation of Ketones by Direct Reaction of Grignard Reagents with Acid Chlorides in Tetrahydrofuran," *Tetrahedron Lett.*, 1979, 4303-4306.

[2] D. Milstein and J. K. Stille, "Mild, Selective, General Method of Ketone Synthesis from Acid Chlorides and Organotin Compounds Catalyzed by Palladium," *J. Org. Chem.*, 44, 1613-1618 (1979).

[3] D. Comins and A. I. Meyers, "A Facile and Efficient Formylation of Grignard Reagents," *Synthesis*, 1978, 403.

[4] H. C. Brown, C. G. Rao, and S. U. Kulkarni, "A Convenient Conversion of Carboxylic Acids into Aldehydes," *Synthesis*, 1979, 704-705.

[5] H. C. Brown, S. U. Kulkarni, and C. G. Rao, "Facile Oxidation of Trialkyl Borates to Aldehydes and Ketones with Pyridinium Chlorochromate. A Remarkable Oxidizability of Such Borate Esters as Compared to Acetates," *Synthesis*, 1979, 702-704.

[6]E. J. Corey and J. W. Suggs, "Pyridinium Chlorochromate. An Efficient Reagent for Oxidation of Primary and Secondary Alcohols to Carbonyl Compounds," *Tetrahedron Lett.*, **1975**, 2647-2650; J. San Filippo, Jr., and C.-I. Chern, "Chromyl Chloride as a Selective Oxidant," *J. Org. Chem.*, **42**, 2182 (1977); S. L. Huang, K. Omura, and D. Swern, "Further Studies on the Oxidation of Alcohols to Carbonyl Compounds by Dimethyl Sulfoxide/Trifluoroacetic Anhydride," *Synthesis*, **1978**, 297; H. Firouzabadi and E. Ghaderi, "Barium Manganate. An Efficient Oxidizing Reagent for Oxidation of Primary and Secondary Alcohols to Carbonyl Compounds," *Tetrahedron Lett.*, **1978**, 839-840.

[7]H. C. Brown, J. L. Hubbard, and K. Smith, "Carbonylation of Organoboranes in the Presence of Triisopropoxyborohydride; A Superior Reagent for the Synthesis of Aldehydes or Primary Alcohols from Alkenes via Hydroboration-Carbonylation," *Synthesis*, **1979**, 701-702.

[8]R. H. Grubbs, "Hybrid-Phase Catalysts," *CHEMTECH*, **1977**, 512-518.

FORMALDEHYDE-[13]C

$$*CH_3OH \xrightarrow[MoO_3-Fe_2O_3]{O_2} H*CHO$$

Procedure[1]

Methanol is catalytically oxidized in a simplified version of the apparatus for the preparation of formaldehyde-^{14}C that has been previously described.[2] The molybdenum oxide-ferric oxide catalyst is prepared by oxidation in a quartz tube that is heated by an electric furnace at 540° in an air stream (140 ml/min) for 2 hr and then at 380-390° for 0.5 hr. Methanol-[13]C (0.50 g, 15 mmol) from a reservoir attached to the quartz reaction tube is carried by the air stream over the catalyst at 380-385° during 20-30 min. The exit gases are passed through acetic acid (50 ml), and the liquid that condenses at the cold end of the reaction tube is washed into the trap with acetic acid. Analysis of the solution indicates a yield of 63% (Note 1).

Notes

1   Analysis is by formation of the dimedone derivative. The solution of product is used directly in the next step.

Other Preparations

Formaldehyde-[17]O and formaldehyde-^{18}O have been prepared in high yield by mercury-lamp irradiation of a mixture of the isotopic carbon monoxide and hydrogen circulating in a quartz apparatus in the presence of mercury vapor.[3] Hydrogenation of potassium cyanide-^{13}C in aqueous solution with palladium-barium sulfate catalyst gives formaldehyde-^{13}C in unspecified yield.[4]

References

[1]A. R. Battersby, E. Hunt, E. McDonald, and J. Moron, "Biosynthesis of Porphyrins and Related Macrocycles. Part II. Synthesis of δ-Amino[5-^{13}C]laevulinic Acid and [11-^{13}C]Porphobilinogen: Incorporation of the Latter into Protoporphyrin-IX," *J. Chem. Soc. Perkin I*, 1973, 2917-2922.

[2]A. Murray and D. L. Williams, *Organic Syntheses with Isotopes*, Interscience, New York, 1958, p. 609.

[3]J. Marling, "Isotope Separation of Oxygen-17, Oxygen-18, Carbon-13, and Deuterium by Ion Laser Induced Formaldehyde Predissociation," *J. Chem. Phys.*, 66, 4200-4225 (1977).

[4]A. Serianni, E. L. Clark, and R. Barker, "Carbon-13-Enriched Carbohydrates. Preparation of Erythrose, Threose, Glyceraldehyde, and Glycolaldehyde with ^{13}C-Enrichment in Various Carbon Atoms," *Carbohydr. Res.*, 72, 79-91 (1979).

2-PROPANONE-2-^{13}C
(^{13}C$_1$-ACETONE)

$$2CH_3*COOLi \xrightarrow{500°} CH_3*COCH_3 + Li_2*CO_3$$

Procedure[1]

Lithium acetate-1-^{13}C (16.7 g, 0.25 mol) (Note 1) contained in a high-temperature-glass boat within a quartz tube is heated over 1 hr to 500°. The evolved product in a gentle stream of nitrogen is condensed in two traps at −78° (8.7 g, 60% yield) and purified by redistillation (Note 2).

Notes

1 The lithium acetate is prepared by treating acetic-1-^{13}C acid with lithium hydroxide (saturated, filtered solution, 10% molar excess), carefully evaporating the solution to dryness in a porcelain dish on an asbestos pad with a burner, and finally drying at 180° for 3 hr.

2 A small amount of carbon dioxide is recovered in a barium hydroxide trap at the end of the reaction apparatus. The lithium carbonate residue is treated with acid, and the carbon-^{13}C dioxide is recovered as barium carbonate-^{13}C. Application of the pyrolysis method to lithium acetate-2-^{13}C, for preparation of 2-propanone-1,3-^{13}C$_2$, would not require recovery of by-product carbon dioxide.

References

[1]B. Stuetzel, W. Ritter, and K-F. Elgert, "^{13}C-Enriched Monomers for NMR-Spectroscopy of Polymers," *Angew. Makromol. Chem.*, 50, 21-41 (1976).

4,4-DIMETHOXY-2-BUTANONE-2-^{13}C

Method I

$$CH_3*COONa \xrightarrow[120-180°]{C_6H_4(COCl)_2} CH_3*COCl \xrightarrow[AlCl_3, CH_2Cl_2]{HC≡CH}$$

$$CH_3*COCH=CHCl \xrightarrow[MeOH]{NaOMe} CH_3*COCH_2CH(OMe)_2$$

Procedure[1]

4-Chloro-3-buten-2-one-2-^{13}C A mixture of sodium acetate-1-^{13}C (3.14 g, 37 mmol) and phthaloyl chloride (20 ml, 0.14 mol) is heated at 120° for 30 min and then at 180° for 3 hr while the acetyl-1-^{13}C chloride is collected in an ice-cooled receiver (2.12 g, 72% yield). The acid chloride (26.6 mmol) in ice-cold methylene chloride (25 ml) is treated with aluminum chloride (3.60 g, 27 mmol) and stirred under an atmosphere of acetylene, initially at 0-5°, then at ambient temperature for 18 hr (Note 1). The dark-red mixture is poured over ice and extracted with ether (2 × 100 ml). The extract is dried over magnesium sulfate and evaporated to a dark oil that is distilled to give the product, bp 50° at 35-45 mm Hg (1.90 g, 67% yield).

4,4-Dimethoxy-2-butanone-2-^{13}C A solution of the chlorovinyl ketone (18 mmol) in methanol (10 ml) at 0° is treated with methanolic sodium methoxide (2.5%, 40 ml) over 15 min. The reaction is quenched with brine (50 ml) and extracted with ether (3 × 50 ml). The extract is dried, evaporated, and distilled to give the product (1.76 g, 73% yield) (Notes 2 and 3).

Method II

$$CH_3*COCH_3 + HCOOMe \xrightarrow[2.\ MeOH,\ HCl]{1.\ NaOMe,\ Et_2O} CH_3*COCH_2CH(OMe)_2$$

Procedure[2]

A mixture of 2-propanone-2-^{13}C (0.50 g, 8.5 mmol) and methyl formate (0.62 g, 10.3 mmol) in ether (2.0 ml) is added dropwise to a stirred suspension of sodium methoxide (0.47 g, 8.6 mmol) in ether (5.0 ml) under nitrogen. The reaction is refluxed 1 hr, cooled, and evaporated carefully at reduced pressure without stirring. The residue is treated with methanol (2.06 ml, 51.5 mmol), stirred for 3 min, treated with methanolic hydrogen chloride (12.7 M, 1.37 ml, 17.4 mmol), and stirred 2 hr at room temperature. The pH is adjusted to 8 with methanolic potassium hydroxide, ether is added, and the mixture is filtered and evaporated at reduced pressure. The residue is dissolved in ether, filtered, evaporated (Note 4), and subjected to short-path distillation at 0.024 mm Hg with a liquid nitrogen-cooled receiver. The distillate is dissolved in ether, dried over sodium sulfate, treated with Type 4A molecular sieve (Note 5), and evaporated to afford the product (0.48 g, 42% yield) (Note 2).

Notes

1 Approx. 700 ml (31 mmol) of acetylene is consumed.
2 IR and NMR data are given.
3 The overall yield from sodium acetate is 35%.
4 Care is taken to avoid possible loss of product through prolonged evaporation.
5 The molecular sieve removes any residual methanol.

References

[1] J. A. Lawson, W. T. Colwell, J. I. DeGraw, R. H. Peters, R. L. Dehn, and M. Tanabe, "The Synthesis of Singly and Doubly ^{13}C-Labeled Mevalonolactone," *Synthesis*, 1975, 729-730.

[2] D. E. Cane and R. H. Levin, "Application of Carbon-13 Magnetic Resonance to Isoprenoid Biosynthesis. II. Ovalicin and the Use of Doubly Labeled Mevalonate," *J. Am. Chem. Soc.*, 98, 1183-1188 (1976).

5-(BENZYLOXY)-3-PENTEN-2-ONE-3-^{13}C

$$*CH_3N(NO)CONH_2 \xrightarrow[\text{Et}_2\text{O}]{\text{KOH, H}_2\text{O}} *CH_2N_2 \xrightarrow[\text{Et}_3\text{N, Et}_2\text{O}]{\text{AcCl}}$$

$$CH_3CO*CHN_2 \xrightarrow[\substack{\text{2. Ph}_3\text{P, CHCl}_3\text{, Et}_2\text{O} \\ \text{3. NaOH, CHCl}_3}]{\text{1. HCl}} CH_3CO*CH=PPh_3$$

$$\xrightarrow[\text{THF}]{\text{PhCH}_2\text{OCH}_2\text{CHO}} PhCH_2OCH_2CH=*CHCOCH_3$$

Procedure[1]

(2-Oxopropylidene-1-^{13}C)triphenylphosphorane 1-(Methyl-^{13}C)-1-nitrosourea (6.92 g, 67.2 mmol) (Note 1) is added slowly with stirring to an ice-salt-cooled mixture of potassium hydroxide (50%, 60 ml) and ether (120 ml). After 2 hr the ether phase is separated, the aqueous phase is extracted with ether (3 × 30 ml), the ether extract is dried over potassium hydroxide and decanted. The solution of diazomethane-^{13}C is treated with triethylamine (6.77 g, 67.6 mmol) and then dropwise with acetyl chloride (5.25 g, 67.3 mmol) in ether (20 ml) while the temperature is kept below $-5°$. The mixture is allowed to stand overnight at $0°$ and then treated with anhydrous hydrogen chloride. The precipitated triethylamine hydrochloride is filtered and washed with ether. The ether solution is evaporated, and the residue is dissolved in chloroform (50 ml), treated with triphenylphosphine (18 g, 68.7 mmol), and refluxed 45 min. The cooled solution is added to ether (600 ml) to precipitate the (2-oxopropyl-1-^{13}C)triphenylphosphonium chloride. A solution of the salt

(6.65 g, 18.8 mmol) in chloroform (40 ml) is shaken with sodium hydroxide (2 N, 9.4 ml) for 15 min, the aqueous phase is extracted with chloroform (3 × 10 ml), and the combined organic extract is dried over sodium sulfate and evaporated to give the phosphorane as a white powder (5.6 g, 30% yield from the nitrosourea) (Note 2).

5-(Benzyloxy)-3-penten-2-one-3-^{13}C A solution of benzyloxyacetaldehyde (8.8 g, 58.7 mmol) and the phosphorane above (14.3 g, 50 mmol) in tetrahydrofuran (230 ml) is refluxed under nitrogen for 40 hr. The residue after the solvent is evaporated is purified by column chromatography (Note 3) (Kieselgel, ether) followed by vacuum distillation, bp 137° at 0.1 mm Hg (6.9 g, 78% yield) (Note 2).

Notes

1 The 1-(methyl-^{13}C)-1-nitrosourea is prepared from methylamine-^{13}C hydrochloride (5.75 g, 85.2 mmol) in 79% yield.
2 ^{13}C-NMR data are given.
3 Chromatography removes triphenylphosphine oxide.

References

[1]A. Gossauer and K. Suhl, "Total Synthesis of Verrucarin E. Its Application to Preparation of a ^{13}C-Labeled Derivative," *Helv. Chim. Acta*, 59, 1698-1704 (1976).

CYCLOPENTANONE-^{18}O

$$H^*O^*OC(CH_2)_4C^*O^*OH \xrightarrow[290°]{Ba(OH)_2} \bigcirc\!\!=\!\!{}^*O + H_2{}^*O + C^*O_2$$

Procedure[1]

Adipic-^{18}O$_4$ acid (6.0 g, 40 mmol) and barium hyroxide (0.25 g, 1.5 mmol) are placed in a 25-ml flask that is attached through an air condenser to a vacuum line having a U-tube trap at −78° for collection of the liquid reaction products and a 1-liter flask with a stopcock for the carbon dioxide-^{18}O$_2$. The apparatus is evacuated, and the reaction flask is heated gradually to 290° in a Wood's metal bath. After about an hour carbon dioxide evolution ceases (Note 1), heating is interrupted, the carbon dioxide is condensed, the ketone is transferred to the trap, and heating is resumed. The process is repeated until the reaction is complete. Most of the water is removed from the trap with a pipet, and the ketone is dried on the vacuum line by distillation through Drierite (3.0 ml, 90% yield).

Notes

1 The reaction stops owing to the lowered temperature caused by the presence of the ketone.

Other Preparations

The synthesis developed for cyclopentanone-3,3,4,4-d_4[2] should be applicable to other isotope isomers, such as cyclopentanone-1-^{13}C and cycopentanone-^{18}O. In this procedure the di-Grignard reagent from 1,4-dibromobutane is treated with carbon dioxide to afford the product in 40% yield. Even if increased yields are not realized with isotopic carbon dioxides, the simplicity and nature of the one-step reaction make it appear attractive and efficient.

References

[1]T.-S. Chen, J. C. Stephens, and L. C. Leitch, "Synthesis of 3-Deuteriocyclopentanone and Cyclopentanone-^{18}O," *J. Labelled Compds.*, 6, 174-178 (1970).

[2]E. W. Della and H. K. Patney, "Synthesis of Octadeuterocyclopentanone," *J. Labelled Compds.*, 9, 651-659 (1973).

CYCLOHEXANONE-2-^{13}C

Procedure[1]

1-(Aminomethyl-^{13}C)cyclopentanol (3.25 g, 28 mmol) (Note 1) in acetic acid (25%, 50 ml) at $-5°$ is treated slowly with sodium nitrite (4 g, 60 mmol) in water (10 ml) and then stirred at 50° for 90 min. The mixture is cooled in ice, saturated with sodium bicarbonate, and extracted with ether. The extract, dried over sodium sulfate, contains the crude product (1.64 g, 66% yield) (Note 2), which is freed of ether by distillation (bath temperature 60°) with a 20-cm Vigreux column.

Notes

1   The starting material contains a trace of ether and a small amount of cyclopentanol.

2   Cyclopentanol (0.039 g) and another impurity, presumably the epoxide (0.062 g), are also present. The product is used directly in the next step.

References

[1]F. Geiss and G. Blech, "Synthesis of o-Terphenyl-2-^{13}C," *J. Labelled Compds.*, 4, 119-126 (1968).

BENZALDEHYDE-^{13}C

Method I

$$Ph*COCl \xrightarrow[\text{Diglyme, } -60°]{\text{Li(t-BuO)}_3\text{AlH}} Ph*CHO$$

Procedure[1]

Benzoyl-^{13}C chloride (1.13 g, 8.0 mmol) in dyglyme (15 ml) at −60° is stirred and treated dropwise with a solution of lithium tri-*tert*-butoxyaluminum hydride (2.03 g, 8.0 mmol) in diglyme (10 ml) (Note 1). The mixture is allowed to warm to room temperature over 1 hr, poured onto ice, filtered, and extracted with ether (3 × 25 ml). The extract is washed with water, dried over sodium sulfate, and distilled in vacuum, bp 27-28° at 0.2 mm Hg (Note 2). The distillate is purified through the bisulfite addition product and redistillation to afford the product (0.509 g, 60% yield).

Method II

$$Ph*COCl \xrightarrow[\text{Quinoline-S}]{\text{H}_2, \text{Pd/C}} Ph*CHO$$

Procedure[2]

A mixture of benzoyl-^{13}C chloride (14 g, 0.10 mol), sodium acetate (25 g, 0.3 mol), palladium-on-charcoal (10%, 3.0 g), and quinoline-sulfur (1.5 ml) in benzene (200 ml) is hydrogenated at 50 psi and ambient temperature for 18 hr (Note 3). The mixture is filtered through Celite, washed with sodium carbonate (5%, 25 ml) and with water (25 ml), and dried over sodium sulfate. The products from five such reactions are combined and distilled at reduced pressure to afford the product, bp 76° at 23 mm Hg (30 g, 56% yield).

Notes

1 Addition is at a rate such that the temperature variation is no more than 2°.
2 Gas chromatographic analysis shows two peaks.
3 A Parr hydrogenation apparatus is used.

References

[1]G. A. Braden and U. Hollstein, "Synthesis of 1-Phenyl-1-^{13}C-ethene-1-^{13}C (*trans*-Stilbene) and Derivatives," *J. Labelled Compds.*, 12, 507-516 (1976).
[2]J. A. Kepler, J. W. Lytle, and G. F. Taylor, "Synthesis of 5,5-Diphenylhydantoin-2,4,5-^{13}C$_3$," *J. Labelled Compds.*, 10, 683-687 (1974).

3,4-(METHYLENEDIOXY)BENZALDEHYDE-^{13}C

Procedure[1]

A solution of 3,4-(methylenedioxy)benzyl-α-^{13}C alcohol (4.4 g, 29 mmol) in pyridine (17 ml) is added dropwise to chromium trioxide-pyridine complex [prepared from chromium trioxide (8. 0 g, 80 mmol) and pyridine (135 ml)] with stirring in an ice bath, and the mixture is allowed to stand for 18 hr at room temperature. The mixture is poured into ice water (675 ml) and extracted with ether. The extract is washed with sulfuric acid (6 N) and sodium chloride solution, dried over sodium sulfate, and evaporated to leave colorless crystals, mp 36-37° (3.9 g, 90% yield).

References

[1]T. Nagasaki, Y. Katsuyama, and H. Minato, "Synthesis of Radioactive and Stable Isotope-Labelled 1-Ethyl-6,7-methylenedioxy-4(1H)-oxocinnoline-3-carboxylic Acids (Cinoxacin)," *J. Labelled Compds.*, **12**, 409-427 (1976).

3,4-BIS(BENZYLOXY)BENZALDEHYDE-^{13}C

Procedure[1]

3,4-Bis(benzyloxy)benzyl-α-^{13}C alcohol (3.8 g, 12.0 mmol) (Note 1) and chromium trioxide-graphite (Seloxcette, 55% CrO_3, 5.0 g, 27.5 mmol) are refluxed in toluene (75 ml) with stirring for 72 hr under nitrogen. The mixture is twice filtered through magnesium sulfate, the solvent is evaporated, and the residue is sublimed at 120° and 0.05 mm Hg to give a white solid, mp 84-86° (3.1 g, 81% yield) (Note 2).

Notes

1 The crude product obtained by lithium aluminum hydride reduction of the acid is used.

2 NMR and MS data are given.

References

[1]M. W. Ames and N. Castagnoli, Jr., "The Synthesis of ^{13}C-Enriched α-Methyldopa," *J. Labelled Compds.*, 10, 195-205 (1974).

3-BROMOBENZALDEHYDE-^{18}O

Procedure[1]

3-Bromobenzylideneaniline (26.5 g, 0.101 mol) is treated with a mixture of water-^{18}O (5 g, 0.278 mol) and sulfuric acid (9.8 g, 0.10 mol) (Note 1). The mixture is filtered, the filtrate is extracted with benzene (50 ml), and the extract is dried over magnesium sulfate and evaporated at reduced pressure. The residual yellow-brown liquid is distilled at reduced pressure to give the product, bp 80.5° at 5 mm Hg (15.2 g, 81% yield, Note 2).

Notes

1   The reaction is slightly exothermic and anilinium sulfate precipitates.
2   The yield is based on organic starting material.

References

[1]M. Kobayashi, H. Minato, and Y. Ogi, "Scrambling of Oxygen in the Diester Produced from Labeled Diacyl Peroxide-Olefin Reaction," *Bull. Chem. Soc. Jap.*, 43, 1158-1160 (1970).

2-DIAZOACETOPHENONE-2-^{13}C

Procedure[1]

1-(Methyl-^{13}C)-1-nitrosourea   A solution of methylamine-^{13}C hydrochloride (500 mg, 7.40 mmol) and potassium cyanate (750 mg, 9.26 mmol) in water (2 ml) is boiled under reflux for 25 min, cooled, and treated with sodium nitrite (500 mg, 7.25 mmol). The solution is cooled to 0° and added during 20 min with stirring to a solution of sulfuric acid (conc., 0.5 ml) in water (3 ml) at $-10°$. The precipitated product is filtered and dried (598 mg, 83% yield) (Note 1).

2-Diazoacetophenone-2-^{13}C   1-(Methyl-^{13}C)-1-nitrosourea (615 mg, 5.97 mmol) is added slowly with magnetic stirring to a mixture of ether (10 ml) and potassium hydroxide (50%, 5 ml) cooled to $-10°$ in an ice-salt bath, and stirring is continued 2 hr at $-5°$. The ether layer is decanted onto potassium hydroxide pellets cooled to $-5°$; the aqueous layer is extracted with ether (4 × 2.5 ml), and the washings are combined with the major portion. The ether solution of diazomethane-^{13}C is decanted into a precooled flask in an ice-salt bath, and triethylamine (420 mg, 4.16 mmol) is added to the stirred solution. A solution of benzoyl chloride (567 mg, 4.04 mmol) in ether (2 ml) is added dropwise with the temperature of the mixture maintained at $-5°$, and stirring is continued for 4 hr at below $0°$. After filtration the solution is evaporated to give the crude product (572 mg), which is recrystalized from hexane at $-20°$, mp 46-48° (356 mg, 60% yield based on benzoyl chloride). A second crop is obtained, mp 41.5-44.5° (85 mg, 75% total yield) (Note 2).

Notes

1   The aqueous filtrate is extracted with dichloromethane, and the extract is washed with saturated sodium chloride, dried, and evaporated to give additional product (32 mg, 85% total yield).

2   Efficient use of diazomethane and other considerations in preparation of labeled aliphatic diazo ketones have been discussed.[2]

References

[1]P. Yates and M. J. Betts, "Base-Induced Rearrangement of γ-Diketones. II. Demonstration of the Occurrence of Skeletal Rearrangement and of the Reversibility of the Reaction," *J. Am. Chem. Soc.*, 94, 1965-1970 (1972).

[2]L. T. Scott and M. A. Minton, "Aliphatic Diazo Ketones. A Modified Synthesis Requiring Minimal Diazomethane," *J. Org. Chem.*, 42, 3757-3758 (1977).

2'-AMINO-4',5'-(METHYLENEDIOXY)ACETOPHENONE-1,2-$^{13}C_2$

Procedure[1]

3′,4′-(Methylenedioxy)acetophenone-1,2-$^{13}C_2$ A solution of α-(methyl-^{13}C)-3,4-(methylenedioxy)benzyl-α-^{13}C alcohol (4.2 g, 25 mmol) in pyridine (10 ml) is added dropwise to chromium trioxide-pyridine complex [from chromium trioxide (10 g, 0.10 mol) and pyridine (140 ml)] with stirring in an ice bath and left at room temperature for 18 hr. The mixture is poured into ice water (600 ml) and extracted with ether. The extract is washed with sulfuric acid (6 N) and water, dried over sodium sulfate, and evaporated. The residue is recrystallized from dichloromethane-ether to give colorless prisms, mp 85-86° (3.9 g, 94% yield).

4′,5′-(Methylenedioxy)-2′-nitroacetophenone-1,2-$^{13}C_2$ The product above (23.8 mmol) is added with stirring in small portions to nitric acid (d 1.42, 40 ml) in an ice bath and stirred for 1 hr. The mixture is poured into ice water (300 ml) and extracted with dichloromethane. The extract is washed with water, dried over sodium sulfate, and evaporated. The residue is recrystallized from dichloromethane-ether to give colorless prisms, mp 123-124° (3.5 g, 70% yield).

2′-Amino-4′,5′-(methylenedioxy)acetophenone-1,2-$^{13}C_2$ The nitro compound above (16.6 mmol) with platinum dioxide (0.85 g) in ethanol (320 ml) is hydrogenated at 13° until 1460 ml of hydrogen is absorbed. The mixture is filtered and evaporated, and the residue is recrystallized from dichloromethane-ether to give colorless needles, mp 169-171° (2.7 g, 90% yield).

References

[1]T. Nagasaki, Y. Katsuyama, and H. Minato, "Synthesis of Radioactive and Stable Isotope-Labelled 1-Ethyl-6,7-methylenedioxy-4(1H)-oxocinnoline-3-carboxylic Acids (Cinoxacin)," *J. Labelled Compds.*, **12**, 409-427 (1976).

2-AMINO-3′,4′-DIHYDROXYACETOPHENONE-^{18}O (NORADRENALONE-^{18}O)

Procedure[1]

Noradrenalone hydrochloride (200 mg, 1.0 mmol) is dissolved in a mixture of methanol (8 ml), acetonitrile (1 ml), and water-^{18}O (0.5 g, 27 mmol). Hydrogen chloride (approx. 3 ml) is bubbled into the solution. Exchange is complete after 24 hr (Note 1).

Notes

1   Aliquots are taken and evaporated to dryness in capillary tubes for mass spectral analysis at 1.5, 7, and 24 hr.

Other Preparations

Owing to the rapid exchange of the noradrenalone with water, conversion to nore-pinephrine-^{18}O by catalytic reduction of the N-benzyl-protected derivative, a procedure successfully used to prepare norepinephrine-d_3, would have to be carried out in water-^{18}O, or be reduced under anhydrous conditions with a metal hydride.[1]

References

[1]R. C. Murphy, "Synthesis of Stable Isotope Labeled Norepinephrine," *J. Labelled Compds.*, **11**, 341-347 (1975).

1-PHENYL-2-PROPANONE-$^{13}C_3$

$$Ph^*CH_2^*COOH \xrightarrow[\substack{2.\ ^*CH_3Li \\ 3.\ H_2O,\,H^+}]{1.\ BuLi,\,Et_2O} Ph^*CH_2^*CO^*CH_3$$

Procedure[1]

Methyllithium-^{13}C is prepared from methyl-^{13}C iodide (15.0 g, 0.11 mol) and lithium (2.5 g, 0.36 mol) in ether (200 ml) to give approx. 180 ml of a 0.45 M solution (73% yield). Phenylacetic-$^{13}C_2$ acid (6.5 g, 48 mmol) in ether (300 ml) in a flask equipped with stirrer, dropping funnel, and drying tube is treated with butyl-lithium (2.6 M, 18 ml, 47 mmol) in ether (30 ml), followed by the above-mentioned solution of methyllithium-^{13}C (81 mmol). The addition is made during 1 hr, and stirring is continued overnight. The mixture is pipetted into ice water containing hydrochloric acid (conc., 20 ml) and extracted with ether. The extract is washed with sodium bicarbonate solution and water, dried over magnesium sulfate, and evaporated to give the crude ketone (6.0 g), which is distilled to afford the product, bp 125° at 45 mm Hg (4.0 g, 60% yield).

References

[1]M. Pomerantz and R. Fink, "Synthesis of Diethyl 1-[(^{13}C)Methyl]-3-phenyl(1,3-$^{13}C_2$)bicyclo-[1.1.0]butane-exo,exo-2,4-dicarboxylate," *J. Labelled Compds.*, **16**, 275-286 (1979).

1-[3,4-BIS(BENZYLOXY)PHENYL]-2-PROPANONE-1-^{13}C

PhCH$_2$O—⟨○⟩—*CHO $\xrightarrow[\text{NH}_4\text{OAc}]{\text{CH}_3\text{CH}_2\text{NO}_2}$ PhCH$_2$O—⟨○⟩—*CH=C(NO$_2$)CH$_3$
PhCH$_2$O—   PhCH$_2$O—

$\xrightarrow[\text{2. H}_2\text{O}]{\text{1. Fe, HOAc}}$ PhCH$_2$O—⟨○⟩—*CH$_2$COCH$_3$
                                                         PhCH$_2$O—

Procedure[1]

1-[3,4-Bis(benzyloxy)phenyl]-2-nitropropene-1-^{13}C   A mixture of 3,4-bis(benzyl-oxy)benzaldehyde-^{13}C (10.0 g, 31 mmol), ammonium acetate (1.2 g, 16 mmol), and nitroethane (85 ml) is refluxed under nitrogen with stirring for 18 hr. From the cooled reaction mixture is obtained the crude product, mp 107-109° (9.64 g, 83% yield) (Note 1). From the concentrated filtrate is obtained a second crop (0.5 g, 5%).

1-[3,4-Bis(benzyloxy)phenyl]-2-propanone-1-^{13}C   A mixture of iron filings (20 mesh, 12.6 g, 0.224 mol) and acetic acid (40 ml) is heated at reflux under nitrogen with vigorous stirring until a grey, milky color is observed (30 min), and the crude nitrostyrene above (9.5 g, 25 mmol) is added. The mixture is heated and stirred an additional 2 hr and filtered hot through Celite, which is washed with hot acetic acid (300 ml). Water (300 ml) is added to the filtrate, and the orange, milky solution is extracted with methylene chloride (3 × 100 ml). The extract is washed with sodium bicarbonate (5%, 3 × 100 ml) and water (100 ml), dried over magnesium sulfate, and evaporated to give the crude product (8.58 g), which on molecular distillation at 160° and 0.05 mm Hg gives a pale-yellow oil (7.5 g, 86% yield) (Note 1).

Notes

1   NMR and MS data are given.

References

[1]M. W. Ames and N. Castagnoli, Jr., "The Synthesis of α-Methyldopa," *J. Labelled Compds.*, **10**, 195-205 (1974).

5,5-DIMETHYL-1,3,3-TRIPHENYL-1,4-HEXANEDIONE-2-^{13}C

PhCO*CHN$_2$ $\xrightarrow[\text{Et}_2\text{O}]{\text{Ph}_2\text{C=C=O}}$

Ph
⟨structure⟩—Ph
Ph—⟨structure⟩=O
        O

$\xrightarrow[\text{2. HCl}]{\text{1. }t\text{-BuLi, Et}_2\text{O}}$

PhCO*CH$_2$CPh$_2$COC(CH$_3$)$_3$ $\xrightarrow[\text{Et}_2\text{O}]{\text{NaOMe}}$ PhCOCPh$_2$*CH$_2$COC(CH$_3$)$_3$

Procedure[1]

4-Hydroxy-2,2,4-triphenyl-3-butenoic-3-^{13}C acid lactone 2-Diazoacetophenone-2-^{13}C (415 mg, 2.84 mmol) in a nitrogen-flushed flask fitted with a magnetic stirrer, serum cap, and reflux condenser is dissolved in ether (5 ml) added through the serum cap, and then treated with diphenyl ketene (600 mg, 3.09 mmol), which is washed in with ether. The mixture is stirred at room temperature for 3 hr, the ether is evaporated in a stream of nitrogen, and the residue is heated on a steam bath for 2 hr. The crude product is chromatographed on a silica column (Note 1) and recrystallized from methanol to give colorless crystals, mp 117-117.5° (540 mg, 61% yield).

5,5-Dimethyl-1,3,3-triphenyl-1,4-hexanedione-2-^{13}C To the lactone above (254 mg, 0.81 mmol) in ether (15 ml) at −15° and under nitrogen is added *tert*-butyllithium in pentane (2.26 *M*, 0.40 ml, 0.90 mmol). The mixture is stirred for 2 min and quenched with hydrochloric acid (3 *N*, 6 ml). The ether layer is washed with sodium bicarbonate solution and water, dried, and evaporated. The residue (218 mg) is crystallized from methanol to give needles, mp 130-130.5° (218 mg, 73% yield) (Note 2).

5,5-Dimethyl-1,2,2-triphenyl-1,4-hexanedione-3-^{13}C The dione above (126 mg, 0.34 mmol) in ether (15 ml) is treated with sodium nydride dispersion in mineral oil (53.8%, 51 mg, 1.14 mmol) and methanol (three drops) for 8 days. The crude product is purified by chromatography (Note 3), and the major component (81 mg) is crystallized from methanol to give colorless needles, mp 128-129° (60 mg, 48% yield) (Note 4).

Notes

1   The product is adsorbed on the column through use of a considerable volume of hot petroleum ether. Elution is conducted with petroleum ether containing increasing amounts of ether; the product (690 mg) elutes with 6-8% ether.
2   An additional 15.5 mg (78% total yield) is obtained from the mother liquors by preparative TLC on silica plates with 10 elutions with 5-6% ether in petroleum ether.
3   Column chromatography on silica with increasing concentration of ether gives a mixture of product and nonrearranged starting material (7:1) that elutes at 20% ether, which is separated by preparative TLC on silica using 5% ether in petroleum ether.
4   In this rearrangement the position of the labeled carbon in the product is unambiguously established by ^{1}H-NMR and shows that the reaction is accompanied by skeletal rearrangement, which allows determination of the mechanism.

References

[1]P. Yates and M. J. Betts, "Base-Induced Rearrangement of γ-Diketones. II. Demonstration of the Occurrence of Skeletal Rearrangement and of the Reversibility of the Reaction," *J. Am. Chem. Soc.*, **94**, 1965-1970 (1972).

2-HYDROXY-1,2-DIPHENYLETHANONE-1,2-$^{13}C_2$
(BENZOIN-α,β-$^{13}C_2$)

$$2Ph*CHO \xrightarrow[\text{EtOH, H}_2\text{O}]{\text{KCN}} Ph*CH(OH)*COPh$$

Procedure[1]

A solution of benzaldehyde-^{13}C (30 g, 0.28 mol) and potassium cyanide (5.0 g, 0.09 mol) in ethanol (95%, 40 ml) and water (32 ml) is refluxed for 1.5 hr, cooled, and filtered. The orange solid is recrystallized from ethanol (95%) to give the product (11.9 g, 40% yield).

References

[1] J. A. Kepler, J. W. Lytle, and G. F. Taylor, "Synthesis of 5,5-Diphenylhydantoin-2,4,5-$^{13}C_3$," *J. Labelled Compds.*, 10, 683-687 (1974).

3,4-DIHYDRO-7-METHOXY-1(2*H*)-NAPHTHALENONE-1-^{13}C
(7-METHOXY-α-TETRALONE-1-^{13}C)

Procedure[1]

4-(4-Methoxyphenyl)butanoic-1-^{13}C acid (2.43 g, 12.5 mmol) is added to polyphosphoric acid (75 g), and the mixture is stirred for 2 hr at 120°. The mixture is chilled, poured onto ice, and extracted with ether. The extract is washed with sodium hydroxide (2 *N*) and water, dried over magnesium sulfate, and evaporated to give the product (1.2 g, 54% yield).

References

[1] G. H. Walker and D. E. Hathway, "Synthesis of N-Phenyl-2-[1,4,5,8-^{14}C] naphthylamine, N-Phenyl-2-[8-^{13}C] naphthylamine, and N-[U-^{14}C] Phenyl-2-naphthylamine," *J. Labelled Compds.*, 12, 199-206 (1976).

1,4-NAPHTHOQUINONE-1-^{13}C OR 1,4-NAPHTHOQUINONE-2-^{13}C

Procedure[1] (Note 1)

Naphthalene-1-^{13}C The Grignard reagent prepared from 1-bromo-3-phenylpropane (12.5 g, 62.5 mmol) and magnesium (1.55 g, 65 mmol) in ether is carboxylated in a vacuum apparatus[2] with carbon-^{13}C dioxide generated from barium carbonate-^{13}C (11.3 g, 57.4 mmol) to give 4-phenylbutyric-1-^{13}C acid, mp 49° (9.5 g, 100% yield).

The acid (9.0 g, 55 mmol) and polyphosphoric acid (75 g) are heated at 80° for 1 hr. The α-tetralone-1-^{13}C is isolated from the reaction and purified by vacuum distillation, bp 116-118° at 9 mm Hg (4.8 g, 60% yield).

The tetralone (2.8 g, 19 mmol) is heated with a mixture of sodium hydroxide and potassium hydroxide (1:1, 0.9 g) for 4 hr at 210-220° to afford naphthalene-1-^{13}C, mp 78-79° (1.74 g, 72% yield), which is recrystallized from alcohol, mp 80-80.2°.

Naphthalene-2-^{13}C The Grignard reagent prepared from 1-bromo-2-phenylethane (12.2 g, 66 mmol) and magnesium (1.6 g, 67 mmol) is carbonated with carbon-^{13}C dioxide generated from barium carbonate-^{13}C (11.7 g, 55.6 mmol) to give 3-phenyl-propionic-1-^{13}C acid (7.1 g, 85% yield).

The acid is reduced with lithium aluminum hydride (2.4 g, 63 mmol) to give 3-phenylpropanol-1-^{13}C, bp 123° at 12 mm Hg (5.0 g, 78% yield).

The alcohol (4.3 g, 32 mmol) is converted to 1-bromo-3-phenylpropane-1-^{13}C, bp 117° at 15 mm Hg (5.6 g, 88% yield). Through the Grignard reaction is obtained 4-phenylbutyric-2-^{13}C acid (1.9 g, 41% yield), which is converted (as above) to naphthalene-2-^{13}C (0.6 g, 40% yield).

1,4-Naphthoquinone-1-^{13}C or 1,4-naphthoquinone-2-^{13}C The oxidation of naphthalene-1-^{13}C or naphthalene-2-^{13}C, as well as mixtures of the two, is accomplished with chromic anhydride in aqueous acetic acid, and the quinones are purified by preparative gas chromatography (Note 2).

Notes

1 Experimental details in this reference are limited.

2 A mass spectrometric method is developed for determination of the extent of labeling in position-1 and position-2 of naphthalene based on the pecularities of the dissociation of 1,4-naphthoquinones by electron impact.

References

[1]V. A. Koptyug, I. S. Isaev, and M. I. Gorfinkel, "The Use of Mass Spectrometry to Determine the Position of the Label in Naphthalene-C^{13}," *Izv. Akad. Nauk. SSR, Ser. Khim.*, 4, 845-849 (1970) (English translation, pp. 794-797).

[2]A. Murray and D. L. Williams, *Synthesis of Organic Preparations with Isotopes of Carbon*, Russian Translation, Vol. 1, IL, 1961, p. 73.

9,10-DIHYDRO-7(8*H*)-BENZO[*a*]PYRENONE-7-^{13}C

Procedure[1]

To 4-(1-pyrenyl)butanoic-1-^{13}C acid (1.01 g, 3.52 mmol) in a 100-ml polyethylene beaker is added hydrogen fluoride (40 ml). The mixture is covered and stirred for 1.5 hr, the hydrogen fluoride is evaporated in a stream of nitrogen, and the solid residue is dissolved in benzene. The solution is filtered, washed with water, dried over sodium sulfate, and evaporated at reduced pressure to give a yellow-brown residue which is crystallized from benzene and recrystallized from xylene to afford yellow plates, mp 173-174° (701 mg, 74% yield).

References

[1]W. P. Duncan, W. C. Perry, and J. F. Engel, "Labeled Metabolites of Polycyclic Aromatic Hydrocarbons. IV. 7-Hydroxybenzo[a]pyrene-7-^{13}C," *J. Labelled Compds.*, 12, 275-280 (1976).

6-BENZO[a]PYRENECARBOXALDEHYDE-^{13}C

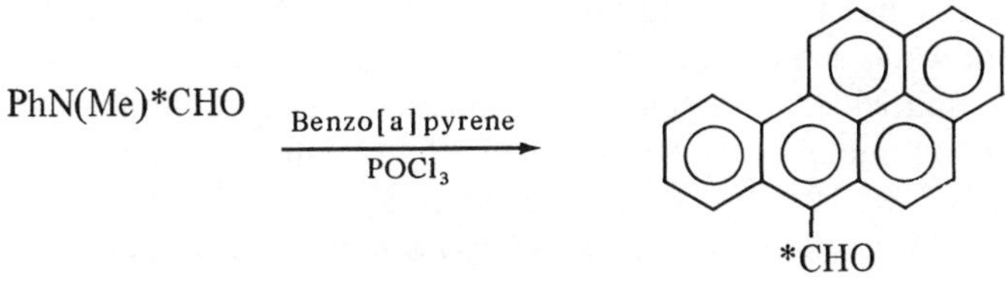

PhN(Me)*CHO $\xrightarrow[\text{POCl}_3]{\text{Benzo[a]pyrene}}$

*CHO

Procedure[1]

Benzo[a]pyrene (4.0 g, 16 mmol), N-methylformanilide-1-^{13}C (4.5 g, 33 mmol), and phosphorus oxychloride (4.5 g, 29 mmol) are heated on a steam bath for 2.5 hr. The solution is cooled, poured into sodium acetate solution (10%, 200 ml), and filtered. The solid is recrystallized twice from chloroform to give the product, mp 202-203° (3.25 g). The residue from the filtrates is chromatographed on alumina with benzene to afford additional product (0.7 g, 88% total yield).

References

[1]R. E. Royer, G. H. Daub, and D. L. Vander Jagt, "Synthesis of Carbon-13 Labelled 6-Substituted Benzo[a]pyrenes," *J. Labelled Compds.*, **12**, 377-380 (1976).

^{13}C$_1$-BENZO[a]PYRENEQUINONES

Method I

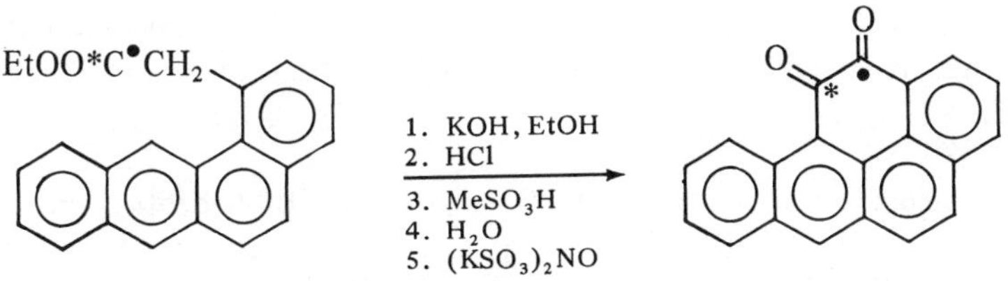

EtOO*C$^\bullet$CH$_2$

1. KOH, EtOH
2. HCl
3. MeSO$_3$H
4. H$_2$O
5. (KSO$_3$)$_2$NO

Procedure[1]

1-Benz[a]anthraceneacetic-α-^{13}C or **1-benz[a]anthraceneacetic-carboxy-^{13}C** acid
To a solution of ethyl 1-benz[a]anthraceneacetate-α-^{13}C or 1-benz[a]anthraceneacetate-carboxy-^{13}C acetate (1.41 g, 4.5 mmol) in ethanol (95%, 50 ml) is added potassium hydroxide (85%, 1.00 g, 15.1 mmol), and the solution is refluxed for 3 hr. Ethanol is removed, and the solid is dissolved in water and acidified with hydrochloric acid to give the crude product as an off-white solid. Recrystallization from benzene gives off-white fibrous needles, mp 202-203° (1.22 g, 95% yield).

11,12-Dihydrobenzo[*a*]pyrene-11,12-dione-11-^{13}C   or   11,12-dihydrobenzo[*a*]pyrene-11,12-dione-12-^{13}C   A solution of the acid above (286 mg, 1.00 mmol) in methanesulfonic acid (10 ml) is stirred under nitrogen for 30 min. The deep-red mixture is poured into water and ice (100 g), and the green precipitate is collected, dissolved in acetone, and added to a solution of dipotassium nitrosodisulfonate (Fremy's salt) (1.07 g, 4.0 mmol) in water (40 ml) buffered with potassium dihydrogen phosphate (0.167 M, 10 ml). The solution is stoppered and shaken on a Parr shaker until it no longer fluoresces under short-wave ultraviolet irradiation. The acetone is evaporated at reduced pressure, and the red-brown precipitate is filtered and dried at reduced pressure, dissolved in methylene chloride, and applied to a chromatography column (silica gel). The product is eluted with chloroform-ethyl acetate (1:1). The solvents are removed, the dark-red solid is dissolved in methylene chloride, and ethyl acetate is added to give dark-red needles, mp 253-254.5° (230 mg, 82% yield).

Method II

Procedure[1]

4-Chryseneacetic-α-^{13}C   or   4-chryseneacetic-*carboxy*-^{13}C   acid   Ethyl 4-chryseneacetate-α-^{13}C or 4-chryseneacetic-*carboxy*-^{13}C is saponified by the procedure in Method I with recrystallization of the product from toluene; mp 205.5-207° (74% yield).

4,5-Dihydrobenzo[*a*]pyrene-4,5-dione-4-^{13}C   or   4,5-dihydrobenzo[*a*]pyrene-4,5-dione-5-^{13}C   A solution of the acid above (0.576 g, 2 mmol) in methanesulfonic acid is stirred under nitrogen for 30 min at 50°. Workup and oxidation with Fremy's salt for 1 hr are carried out as described in Method I. The red-brown precipitate from the oxidation reaction is heated with sodium carbonate (5%) to give the bright orange-red quinone, which is chromatographed (silica gel) and recrystallized from chloroform-ethyl acetate to give red-orange crystals, mp 256-257° (0.48 g, 85% yield).

References

[1]R. S. Bodine, M. Hylarides, G. H. Daub, and D. L. Vander Jagt, "^{13}C-Labeled Benzo[*a*]pyrene and Derivatives. 1. Efficient Pathways to Labeling the 4, 5, 11, and 12 Positions," *J. Org. Chem.*, **43**, 4025-4028 (1978).

4

Alcohols, Ethers, and Phenols

Almost all schemes that have been developed for labeling hydroxy compounds at the site of the functional group involve reduction or condensation of a carbonyl compound or carboxylic acid derivative. Thus the problem, for the most part, depends on the availability of the higher oxidation-state materials, since a wide variety of selective, high-yield catalytic and hydride reduction and organometallic condensation reactions are known.

Direct introduction of the basic isotopic starting material, carbon monoxide, such as through hydroformylation[1] or borane carbonylation[1,2] may find application for specific compounds; for example, with certain tertiary alcohols that are not satisfactorily prepared through the Grignard reaction.[1] Labeling at positions other than the functional group, of course, involves an even wider variety of possible reaction types. Asymmetric syntheses, such as with hydrides[3] or homogeneous hydrogenation catalysts,[4] should receive consideration when chiral labeled alcohols are required. Phase-transfer catalysis[5] and use of trifluoromethanesulfonates[6] in preparation of ethers can be expected to be increasingly useful for isotopic syntheses.

References

[1]C. A. Buehler and D. E. Pearson, *Survey of Organic Syntheses*, Wiley-Interscience, New York, 1970, pp. 190-191.

[2]H. C. Brown, J. L. Hubbard, and K. Smith, "Carbonylation of Organoboranes in the Presence of Triisopropoxyborohydride; A Superior Reagent for the Synthesis of Aldehydes or Primary Alcohols from Alkenes via Hydroboration-Carbonylation," *Synthesis*, 1979, 701-702.

[3]S. Colonna and R. Fornasier, "Asymmetric Induction in the Borohydride Reductions of Carbonyl Compounds by Means of Chiral Phase-transfer Catalysts. Part 2," *J. Chem. Soc. Perkin I*, 1978, 371-373.

[4]M. D. Fryzuk and B. Bosnich, "Asymmetric Synthesis. An Asymmetric Homogeneous Hydrogenation Catalyst Which Breeds Its Own Chirality," *J. Am. Chem. Soc.*, **100**, 5491-5494 (1978).

[5]W. P. Weber and G. W. Gokel, "Phase Transfer Catalysis. Part II: Synthetic Applications," *J. Chem. Educ.*, 55, 429-433 (1978).

[6]R. D. Howells and J. D. McCown, "Trifluoromethanesulfonic Acid and Derivatives," *Chem. Rev.*, 1977, 69-92.

METHANOL-[13]C

Method I

$$*CO_2 + 3H_2 \xrightarrow[200-250°]{Cu-Zn-Cr} *CH_3OH + H_2O$$

Procedure[1,2]

Catalyst A solution of cupric nitrate trihydrate (217 g, 0.90 mol), chromic nitrate nonahydrate (35 g, 0.09 mol), and zinc nitrate hexahydrate (78 g, 0.26 mol) in water (2 liter) is added slowly with vigorous stirring to a solution of sodium carbonate (138 g, 1.29 mol) in water (1.5 liter) in a 4-liter beaker maintained at 75-85°. After cooling, the precipitate is filtered, washed with water, dried, returned to the 4-liter beaker, and slowly heated to 250° until nitrogen oxides are no longer evolved. The cooled brown-black powder is moistened with water to give a thick paste, which is extruded into 1-2-mm-diameter pellets (Note 1), which are dried and transferred to the catalyst chamber of the synthesis unit (Note 2). The apparatus is evacuated and pressured with helium or nitrogen to 100 psig, and the catalyst is heated to approx. 230°. With the heater off hydrogen is added in small increments (1-5 psi) to maintain the temperature at 220-240° until hydrogen uptake is complete. The apparatus is then evacuated while the catalyst is maintained at 220-240° for several hours (Note 3).

Methanol-[13]C To the apparatus suspended in a cooling water bath and containing residual hydrogen pressure is added carbon-^{13}C dioxide (125 psi, 0.25 mol) (Note 4) and hydrogen (approx. 500 psi, 1.0 mol) (Note 5). The catalyst heater is turned on and when operating temperature (220-240°) and convective circulation have been established the pressure begins to decrease. After the pressure has dropped to a constant value, indicating complete reduction of the carbon dioxide (Note 6), the heater is turned off, the apparatus is allowed to cool, and the product is withdrawn to give a mixture of methanol-^{13}C and water in approximately equal molar ammounts (12.2 g, 96% yield) (Note 7). Anhydrous methanol is obtained by bulb-to-bulb vacuum transfer from Type 3A molecular sieve following separation from most of the water by distillation at atmospheric pressure.

Method II

$$*CO_2 \xrightarrow[\text{2. R'OH}]{\text{1. LiAlH}_4\text{, ROR}} *CH_3OH$$

$$[R = C_2H_5OCH_2CH_2\text{-}; \; R' = CH_3(CH_2)_3(OCH_2CH_2)_2\text{-}]$$

Procedure[7]

The carbon-^{13}C dioxide generated from barium carbonate-^{13}C (59.1 g, 0.30 mol) (Note 8) is introduced during 3-4 hr into a magnetically stirred solution of lithium aluminum hydride (18 g, 0.47 mol) (Note 9) in diethyleneglycol diethyl ether (1 liter) (Note 10). To the resulting mixture is added over 2-3 hr with stirring diethyleneglycol monobutyl ether (540 g, 3.3 mol). The product is distilled from the mixture under vacuum at 55-60° into a liquid-nitrogen-cooled receiver. Several repetitions of the distillation-condensation give the anhydrous product (8.4-9.4 g, 85-95% yield), free of high-boiling impurities.

Notes

1   Extrusion of the moist oxide mixture into pellets (more readily accomplished than with the carbonate form) can be carried out with a household garlic press.[3] The yield of oxides, in atomic ratio of 72:21:7 Cu:Zn:Cr, is approx. 100 g (125 cm^3). The amount required depends on the particular apparatus.

2   The apparatus, based on a similar type unit developed for conversion of carbon dioxide to methane for radiocarbon dating,[4] is the same as that used for synthesis of methane-^{13}C,[5] except for a different catalyst and mode of operation. It has been constructed from stainless steel components with 1-liter and 2-liter volumes,[2,5] and presumably would function well in smaller or larger sizes. The catalyst pellets are held around the cylindrical stainless steel heating element by copper screen. The heater must be encased in quartz or copper to prevent excessive production of methane and elemental carbon.[2,5] The mixture of gases circulates through the catalyst by convection up the surrounding chimney, out holes at its top, and down the cooled vessel walls, which also condense the methanol and water. The high heat capacity of hydrogen necessitates adequate heating and cooling, particularly at higher pressures. Another configuration that also uses convective circulation of the gases through the catalyst bed, but of smaller volume and constructed entirely of commercially available components, has been described.[1] A large-scale unit, which utilizes a circulation pump, has also been very effective.[1]

3   Complete conversion of the catalyst is not realized by hydrogen alone; initial reductions of carbon dioxide to methanol are accompanied by the odor of methylamine. This is eliminated during several small, trial runs with nonisotopic carbon dioxide. No change in catalyst activity has been noted over several years of operation.[1,2]

4   The method is based on a procedure for industrial production of methanol from carbon monoxide.[6] The patent indicates the beneficial effect of carbon dioxide on catalyst life; however, reduction of carbon dioxide (only) is not discussed.

5   The pressure-quantity relationship depends on the volume of the particular apparatus.

6   The value of the final pressure depends on the particular apparatus, temperature, and quantity of excess hydrogen. Reaction time and pressure are related exponentially.[1]

7   Analysis is by GC or ^{1}H-NMR, which can also determine the carbon-13 concentration.[1] The molar ratio of methanol to water is not strictly 1:1 for individual preparations owing to holdup, temperature of the condensate, incomplete removal, incomplete reduction of intermediate carbon monoxide, and so on.

8   Formic acid (50%, 120 ml) is added dropwise to the barium carbonate with magnetic stirring; the gas is dried by passage through sulfuric acid and phosphorus pentoxide. A rubber balloon monitors the evolution and uptake of the gas in the closed system.

9   Use of lithium aluminum hydride without purification is given in another investigation[8] as the cause of dilution of the isotope (72% instead of 90%) from methoxide formed by prior absorption of atmospheric carbon dioxide.

10 High boiling reagents simplify isolation of the product.

Other Preparations

Carbon monoxide could also be used in Method I, which was originally developed for this purpose.[6] Hydride reduction, such as Method II, could be applied to formic acid and certain derivatives; however, carbon dioxide is normally the most available starting material. Methanol-^{12}C (99.999 at. % ^{12}C) and methanol-^{13}C-d_4 are also prepared by Method I; the latter compound uses deuterium at relatively high pressure that may not be readily available except in certain government laboratories. Tetrahydrofurfuryl tetrahydropyranyl ether and tetrahydrofurfuryl alcohol have been used in a procedure similar to Method II to prepare methanol-^{13}C in excellent yield.[8,9]

References

[1]D. G. Ott, V. N. Kerr, T. W. Whaley, T. W. Benziger, and R. K. Rohwer, "Synthesis with Stable Isotopes: Methanol-^{13}C, Methanol-^{13}C-D$_4$ and Methanol-^{12}C," *J. Labelled Compds.*, 10, 315-324 (1974).

[2]D. G. Ott and T. G. Sanchez, Los Alamos Scientific Laboratory, Los Alamos, NM, unpublished data, Dec. 1975.

[3]M. J. Ott, Los Alamos, NM, personal communication, April 1976.

[4]R. W. Buddemier, A. Y. Young, A. W. Fairhall, and J. A. Young, "Improved System of Methane Synthesis for Radiocarbon Dating," *Rev. Sci. Instrum.*, 41, 652-656 (1970).

[5]D. G. Ott, V. N. Kerr, T. G. Sanchez, and T. W. Whaley, "Syntheses with Stable Isotopes: Cyanide-^{13}C, Methane-^{13}C, Methane-^{13}C-d_4, and Methane-d_4," *J. Labelled Compds.*, 17, 255-262 (1980).

[6]P. Davies and F. F. Snowdon, "Catalysts for the Reaction of Carbon Monoxide with Hydrogen," U.S. Patent 3,326,956 (June 20, 1967).

[7]B. Stuetzel, W. Ritter, and K-F. Elgert, "^{13}C-Enriched Monomers for NMR Spectroscopy of Polymers," *Angew. Makromol. Chem.*, **50**, 21-41 (1976).

[8]M. Pomerantz and R. Fink, "Synthesis of Diethyl 1-[(^{13}C)Methyl]-3-phenyl(1,3-^{13}C$_2$)bicyclo-[1.1.0]butane-exo,exo-2,4-dicarboxylate," *J. Labelled Compds.*, **16**, 275-286 (1979).

[9]M. A. G. El-Fayoumy, H. C. Dorn, and M. A. Ogliaruso, "Preparation of Trimethyloxosulfonium-^{13}C Iodide," *J. Labelled Compds.*, **13**, 433-436 (1977).

METHANOL-^{18}O

$$H_2{*}O \xrightarrow[\text{Diglyme, HCl}]{(BuO)_3CH} BuOCH{*}O \xrightarrow[\substack{2.\ H_2O,\\ HOCH_2CH_2OH}]{\substack{1.\ LiAlH_4,\\ Diglyme}} CH_3{*}OH$$

Procedure[1]

To a mixture of tributyl orthoformate (7.5 ml, 28.1 mmol), diglyme (5 ml), and water-^{18}O (0.5 ml, 27.6 mmol) is added hydrogen chloride (2.5 ml, approx. 0.1 mmol) from a syringe with stirring. The mixture immediately becomes homogeneous and is distilled through a small Vigreux column. Butyl formate-*carboxyl*-^{18}O and 1-butanol are collected, bp 107-118°, and slowly added to an ice-cooled mixture of lithium aluminum hydride (1.6 g, 42 mmol) in diglyme (40 ml). After 20 min at room temperature the resulting suspension is again cooled in ice and stirred while water (0.8 ml) is slowly dropped in. Ethylene glycol (5 ml) is added (Note 1), and the mixture is distilled through a small Vigreux column to afford the product, bp 64.5-66° (0.76 g, 86% yield) (Note 2).

Notes

1   Use of ethylene glycol to liberate the methanol avoids the problem of separating the product from the butanol-water azeotrope.

2   Methyl-^{18}O formate is obtained by distillation from a refluxing solution of methanol-^{18}O in an eight-fold excess of formic acid (98%). Mass spectrometry of the ester shows the same oxygen-18 concentration in the methoxyl group as in the water-^{18}O.

Other Preparations

Methanol-^{18}O has been prepared by heating sodium hydroxide-^{18}O and methyl iodide in a Carius tube at 90° for 5 days.[2]

References

[1]C. B. Sawyer, "A Simple High Yield Synthesis of Methanol-^{18}O and Ethanol-^{18}O," *J. Org. Chem.*, **37**, 4225-4226, (1972).

[2]A. P. Cox and S. Waring, "Preparation of Methyl Nitrate and Nitromethane Labelled with Nitrogen-15 and Oxygen-18," *J. Labelled Compds.*, 9, 153-157 (1973).

ETHANOL-$^{13}C_2$

$$*CH_3*COOH \xrightarrow[Re, H_2O]{H_2} *CH_3*CH_2OH$$

Procedure[1]

A mixture of acetic-$^{13}C_2$ acid (61.8 g, 1.00 mol), water (150 ml), and rhenium heptoxide (2 g) in a 1-liter stainless steel, stirred autoclave (Note 1) is hydrogenated at an initial pressure of 700 psi hydrogen and at 180°. The hydrogen pressure is maintained between 700 and 1000 psi until no further uptake occurs (approx. 7-10 days) (Note 2), the vessel is cooled, and the contents are transferred and distilled to give the crude product, bp 72-93° at 580 mm Hg (107 g total, 44 g ethanol by GC analysis, 92% yield) (Note 3). Anhydrous material is obtained by treating the aqueous product with Type 3A molecular sieve and subsequent bulb-to-bulb vacuum transfer (Note 4).

Notes

1 A stainless steel gas cylinder containing a magnetic stirring bar, fitted with a valve after the reactants are added, and heated with a mantle or oil bath and heating tape can serve as a reaction vessel.
2 The theoretical amount of hydrogen is absorbed; insufficient reaction affords ethyl-$^{13}C_2$ acetate-$^{13}C_2$ in the product. Too high a reaction temperature leads to lowered yields through ethylene-$^{13}C_2$ production.
3 The hydrogenation is slow; however, yields are excellent, the procedure is simple, and anhydrous starting material is not necessary (indeed, water is required).
4 Physical properties and NMR and IR data are given.

Other Preparations

Ethanol-1-^{13}C and ethanol-2-^{13}C have been prepared by the same procedure.[1]

References

[1]V. N. Kerr and D. G. Ott, "Preparation of D- and L-Alanine-2,3-$^{13}C_2$," *J. Labelled Compds.*, 15, 503-509 (1978).

ETHANOL-^{18}O

$$H_2*O \xrightarrow[HCl]{CH_3CH(OPr)_2} CH_3CH*O$$

$$\xrightarrow[\text{2. H}_2\text{O}]{\substack{\text{1. LiAlH}_4, \\ \text{Diglyme}}} \text{CH}_3\text{CH}_2\text{*OH}$$

Procedure[1]

Acetaldehyde-^{18}O To a stirred mixture of 1,1-dipropoxyethane (3.1 ml, 17.5 mmol) and water-^{18}O (0.25 ml, 12.8 mmol) is added hydrogen chloride (0.8 ml, approx. 0.03 mmol) from a syringe. Slow distillation through a Vigreux column into a receiver at $-78°$ affords the product, bp 21-24° (0.60 g, 95% yield).

Ethanol-^{18}O The aldehyde above (0.56 g, 12 mmol) is slowly added to an ice-cooled mixture of lithium aluminum hydride (0.15 g, 3.9 mmol) in diglyme (10 ml). Water (0.6 ml) is cautiously added, and the product is distilled through a Vigreux column to give ethanol-water azeotrope, bp 78-79° (0.53 g, 90% yield) (Note 1).

Notes

1 Mass spectrometry shows the oxygen-18 concentration in the product to be the same as that in the water-^{18}O.

References

[1]C. B. Sawyer, "A Simple High Yield Synthesis of Methanol-^{18}O and Ethanol-^{18}O," *J. Org. Chem.*, 37, 4225-4226 (1972).

2-AMINOETHANOL-^{15}N

Method I

Procedure[1]

Potassium phthalimide-^{15}N Ammonia-^{15}N, generated from ammonium-^{15}N chloride (2.73 g, 50 mmol), is introduced into a solution at $-10°$ of phthalic anhydride

(7.4 g, 50 mmol) in methanol (120 ml). After 1 hr the solvent is distilled, and the residue is heated to 300°. The light-brown melt solidifies on cooling to afford crude phthalimide-^{15}N, mp 235-238° (7.4 g, 100% yield), which is dissolved in warm alcohol (300 ml) and treated at room temperature with potassium ethoxide (50 mmol) in alcohol (100 ml). The solution, containing precipitated platelets, is evaporated, and the residue is suspended in acetone, filtered, and dried to give the product (9.3 g, 100% yield).

N-(2-Bromoethyl)phthalimide-^{15}N A suspension of the potassium phthalimide (4.65 g, 25 mmol) in acetone (100 ml) is treated with 1,2-dibromoethane (9.3 g, 50 mmol) and stirred at 60° for 40 hr. The mixture is evaporated, and the residue is extracted with warm alcohol (3 × 20 ml). The extract is evaporated at reduced pressure, and the product (4.2 g) is recrystallized from methanol (10 ml), mp 80-82° (3.09 g, 48% yield).

Ethanolamine-^{15}N hydrochloride The bromo compound above (2.3 g, 9 mmol) is heated for 2 hr with sodium hydroxide solution (30%, 30 ml). The cooled solution is brought to pH 1 with hydrochloric acid, and the precipitated phthalic acid is removed by filtration. The filtrate is evaporated at reduced pressure, and the residue is washed with ether. The remaining amine hydrochloride is extracted with ethanol (3 × 15 ml), the extract is evaporated, and the residue is dried under vacuum to give the product, mp 68-70° (0.9 g, 74% yield).

Method II

$$\xrightarrow[\text{2. HCl}]{\text{1. KOH}} \text{HOCH}_2\text{CH}_2\text{*NH}_2\cdot\text{HCl} \xrightarrow{\text{NaOMe}} \text{HOCH}_2\text{CH}_2\text{*NH}_2$$

Procedure[2]

N-(2-Hydroxyethyl)phthalimide-^{15}N Phthalimide-^{15}N (11.54 g, 77.9 mmol) is heated at 200° with ethylene carbonate (8.8 g, 100 mmol) for 2 hr. The cooled solution is treated with boiling water (150 ml) and charcoal and filtered. The white solid is dried at 50° to give the product, mp 115-120° (10.0 g, 66% yield).

2-Aminoethanol-^{15}N The hydroxyethyl compound above (10.0 g, 52 mmol), potassium hydroxide (10 g, 0.18 mol), and water (20 ml) are refluxed for 10 hr. The solution is distilled, finally at 15 mm Hg, into an excess of hydrochloric acid (3 N) cooled in a Dry Ice bath. The aqueous solution is acidified with hydrochloric acid (conc.) to pH 2, evaporated repeatedly at reduced pressure using benzene and ethanol additions (Note 1), and treated with ether to give the hydrochloride as a white, hygroscopic solid (5.0 g). The product (51 mmol) is intimately mixed with sodium methoxide (2.84 g, 52.6 mmol), and the mixture is distilled at a bath temperature of 110° until methanol is no longer collected. Distillation at 90-130° bath temperature and 15 mm Hg into a receiver at −70° affords the product (2.29 g, Note 2).

Notes

1 The evaporations are repeated until the product solidifies on addition of ether.
2 Mass spectral analysis discloses the presence of water and methanol (approx. 15%). The overall yield from ammonium chloride is 34%.

Other Preparations

By Method I *N*-(3-bromopropyl)phthalimide-^{15}N has been prepared (38% yield) and converted to 3-amino-1-propanol-^{15}N hydrochloride.[1] Reduction of ethyl glycinate-^{15}N hydrochloride and ethyl glycinate-1-^{13}C hydrochloride with lithium aluminum hydride has given 2-aminoethanol-^{15}N hydrochloride and 2-aminoethanol-1-^{13}C hydrochloride, which are treated with sulfuric acid to give the sulfate esters that are then converted to aziridine-^{15}N and aziridine-^{13}C$_1$ by treatment with potassium hydroxide at 130°.[3]

References

[1]J. Volford and D. Banfi, "Synthesis of ^{15}N-Labeled 2-Substituted 2-Thiazolines and Analogous Thiazines," *J. Labelled Compds.*, **11**, 419-426 (1975).
[2]W. L. Mendelson, L. E. Weaner, L. A. Petka, and D. W. Blackburn, "The Preparation of ^{15}N-Phenoxybenzamine," *J. Labelled Compds.*, **11**, 349-353 (1975).
[3]B. Bak and S. Skaarup, "Preparation of [2-D], [2,2'-D$_2$], [1-^{15}N], and [2-^{13}C] Enriched Ethylenimines," *J. Labelled Compds.*, **7**, 445-448 (1971).

2-[BENZYL(1-METHYL-2-PHENOXYETHYL)AMINO]ETHANOL-^{15}N

$$H_2\text{*}NCH_2CH_2OH \xrightarrow[Na_2CO_3, 190°]{PhOCH_2CH(CH_3)Cl} PhOCH_2CH(CH_3)\text{*}NHCH_2CH_2OH$$

$$\xrightarrow[NaHCO_3, EtOH]{PhCH_2Cl} PhOCH_2CH(CH_3)\text{*}N(CH_2Ph)CH_2CH_2OH$$

Procedure[1]

2-Aminoethanol-^{15}N (2.2 g, 36 mmol) (Note 1), 2-chloro-1-phenoxypropane (15 ml), and sodium carbonate (5.7 g, 54 mmol) are heated in a bath at 190° for 4 hr. The reaction is quenched with water and extracted with chloroform, which is washed with water and extracted with hydrochloric acid (5%). The acid solution is adjusted to pH 10 with sodium hydroxide (2 *N*), and the free base is extracted with chloroform, which is dried and evaporated to give an oil that is triturated with hexane to afford white, crystalline 2-(1-methyl-2-phenoxyethylamino)ethanol-^{15}N, mp 72-73° (2.12 g, 30% yield) (Note 2).

The amine (1.99 g, 10.2 mmol), benzyl chloride (1.31 g, 10.4 mmol), sodium bicarbonate (0.874 g, 10.4 mmol), and ethanol (20 ml) are refluxed for 24 hr. The mixture is filtered from the inorganic salts, and the alcohol is evaporated at reduced pressure to give an oily residue, which is dissolved in ether, dried over magnesium sulfate, and concentrated to give the product (quantitative yield).

Notes

1 The starting material contains some water and methanol.
2 Commercial ethanolamine has given yields of 59% and 67%; the low yield is attributed to the methanol or water in the starting material.

References

[1]W. L. Mendelson, L. E. Weaner, L. A. Petka, and D. W. Blackburn, "The Preparation of ^{15}N-Phenoxybenzamine," *J. Labelled Compds.*, **11**, 349-353 (1975).

CYCLOPENTANOL-1-^{13}C

$$H*COOCHMe_2 \xrightarrow[\text{THF}]{BrMg(CH_2)_4MgBr} \quad \text{⬠}*\text{-OH}$$

Procedure[1]

A solution of isopropyl formate-^{13}C (22.7 g, 0.23 mol) (Note 1) in tetrahydrofuran (250 ml) is added, slowly over a 3-hr period with stirring, to the Grignard reagent prepared from 1,4-dibromobutane (90 ml, 0.75 mol), magnesium turnings (60 g, 2.5 mol), and tetrahydrofuran (1.8 l). The reaction mixture is hydrolyzed by the slow addition of ammonium sulfate solution (saturated, approx. 0.2 liter) amd water (0.5 liter). The organic layer is separated, dried over potassium carbonate, and evaporated at reduced pressure. Fractionation of the residue affords the product, bp 41-43° at 0.6 mm Hg (12.1 g, 61% yield).

Notes

1 The ester contains approx. 10% isopropyl alcohol.

References

[1]S. D. Larsen, P. J. Vergamini, and T. W. Whaley, "Synthesis of Cyclopentadienyl-X-^{13}C Thallium," *J. Labelled Compds.*, **11**, 325-332 (1975).

1-(AMINOMETHYL-^{13}C)CYCLOPENTANOL

K*CN + [cyclopentane]=O $\xrightarrow{\begin{array}{l}1.\ H_2O,\ Et_2O\\ 2.\ HCOOH\end{array}}$ [cyclopentane ring with *CN and OH]

$\xrightarrow{\begin{array}{l}1.\ Ac_2O\\ 2.\ LiAlH_4,\ Et_2O\end{array}}$ [cyclopentane ring with *CH$_2$NH$_2$ and OH]

Procedure[1]

1-Hydroxycyclopentanecarbonitrile-^{13}C To a mixture of potassium cyanide-^{13}C (2.56 g, 40 mmol), cyclopentanone (6.7 g, 80 mmol), water (8 ml), and ether (30 ml) is added formic acid (90%, 5.3 g, 0.1 mol) during 3 hr at −5°. The reaction is stirred an additional 4 hr at 0° and then allowed to stand in the refrigerator for 2 days. The ether phase is separated, and the aqueous layer is extracted with ether (5 × 25 ml). The combined extracts are dried over sodium sulfate and evaporated to give the product (3.74 g, 85% yield).

1-(Aminomethyl-^{13}C)cyclopentanol The cyanohydrin above (34 mmol) is refluxed 3 hr with acetic anhydride (4 g, 40 mmol) and acetyl chloride (3 drops). The mixture is taken up in ether and neutralized with sodium bicarbonate solution, and the ether solution is dried over sodium sulfate and evaporated. The residue is dissolved in ether and added to a solution of lithium aluminum hydride (10 g, 0.27 mol) in ether (70 ml) at such a rate that boiling is just maintained. Finally, the mixture is refluxed 90 min, cooled, made strongly alkaline with sodium hydroxide (2 N), and filtered. The filtrate and ether washings (800 ml) (containing 2.9 g of crude product) are extracted with hydrochloric acid (15%, 20 ml) and washed with water (6 × 10 ml). The ether-washed acid solution is cooled and saturated with potassium hydroxide. The free amine is extracted with ether, and the solution is dried over potassium carbonate and evaporated to afford the product (3.25 g, 81% yield) (Note 1).

Notes

1 The product, containing a trace of ether and cyclopentanol (3%), is used directly in the next step.

References

[1]F. Geiss and G. Blech, "Synthesis of o-Terphenyl-2-^{13}C," *J. Labelled Compds.*, 4, 119-126 (1968).

1-METHYLCYCLOHEXANOL-1-^{13}C

$$CH_3\text{*}COOEt \xrightarrow[\text{Et}_2O]{\text{BrMg(CH}_2)_5\text{MgBr}}$$

Procedure[1,2]

A solution of ethyl acetate-1-^{13}C (12 g, 13.5 mmol) in ether (150 ml) is slowly added to the ice-cooled, vigorously stirred Grignard reagent prepared from 1,5-dibromopentane (95 g, 0.41 mol), magnesium (21 g, 0.88 mol), and ether (500 ml). The mixture is stirred overnight at room temperature, refluxed 1 hr, and treated with ammonium chloride solution (saturated). The aqueous phase is extracted several times with ether, and the combined extracts are dried over sodium sulfate and distilled to afford the product, bp 65° at 24 mm Hg (12.7 g, 80% yield).

Other Preparations

By the same procedure 1-(methyl-^{13}C)cyclohexanol has been prepared without isolation from ethyl acetate-2-^{13}C.[2]

References

[1]B. Stuetzel, W. Ritter, and K-F. Elgert, "^{13}C-Enriched Monomers for NMR-Spectroscopy of Polymers," Angew. Makromol. Chem., 50, 21-41 (1976).
[2]G. A. Braden and U. Hollstein, "Synthesis of 1-Phenyl-2-phenyl-1-^{13}C-ethene-1-^{13}C (trans-Stilbene) and Derivatives," J. Labelled Compds., 13, 507-516 (1976).

4-NITROPHENOL-1-^{13}C

$$(CH_3)_2\text{*}CO \xrightarrow[\text{2. HCl}]{\substack{\text{1. NaC(NO}_2)(\text{CHO})_2, \\ \text{NaOH}}}$$

Procedure[1]

A solution of sodium nitromalonaldehyde monohydrate (3.31 g, 21.1 mmol) and sodium hydroxide (0.595 g, 14.8 mmol) in water (70 ml) cooled to 4° is treated with 2-propanone-2-^{13}C (1.00 g, 17.2 mmol), diluted with water to 200 ml, and allowed to stand for 5 days at 2°. To the major portion of the solution (175 ml, 87.5%) (Note 1) is added cold sodium hydroxide solution (50%, 85 ml) slowly such that the temperature is maintained at 4-10°. The mixture is stirred 2 hr at

$6°$, acidified with hydrochloric acid ($12\ N$) while the temperature is kept below $30°$, and filtered from a brown solid. For convenience the mixture is divided into two equal portions, and each is extracted with ether (5×200 ml). The extracts are dried over sodium sulfate and concentrated at reduced pressure. The dark yellow-brown residue is dissolved in water (minimum amount), and the cooled solution is treated with cold sodium hydroxide solution (50%) to precipitate the sodium salt of the phenol, which is dissolved in water, acidified, and extracted into ether. The ether solution is dried and evaporated, and the residue is sublimed at reduced pressure to afford the light-yellow crystalline product, mp 108-$109.5°$ (1.54 g, 74% yield) (Note 2).

Notes

1 A 25-ml aliquot worked up by standard procedures affords the product in only 13% yield.
2 The yield is based on amount of starting material in the aliquot.

Other Preparations

By a similar procedure 4-nitrophenol-2,6-$^{13}C_2$ has been prepared in 65% crude yield from 2-propanone-1,3-$^{13}C_2$.[2] Methylation with methyl iodide and potassium hydroxide in dimethyl sulfoxide gives 4-nitroanisole-2,6-$^{13}C_2$ in 91% yield.[2]

References

[1]G. L. Swartz and W. M. Gulick, Jr., "Synthesis of Carbon-13 Enriched Nitrobenzene," *J. Labelled Compds.*, **11**, 525-534 (1975).
[2]V. Viswanatha and V. J. Hruby, "Synthesis of [3',5'-$^{13}C_2$]-Tyrosine and Its Use in the Synthesis of Specifically Labeled Tyrosine Analogues of Oxytocin and Arginine-Vasopressin and Their 2-D-Tyrosine Diastereoisomers," *J. Org. Chem.*, **44**, 2892-2896 (1979).

4-AMINOPHENOL-^{18}O

$$PhNHOH \xrightarrow[H_2SO_4]{H_2{}^{*}O} H^{*}O-\!\!\left\langle\!\bigcirc\!\right\rangle\!\!-NH_2$$

Procedure[1]

A solution of *N*-phenylhydroxylamine (0.43 g, 4.0 mmol) in water-^{18}O (38 ml) containing sulfuric acid (conc., 2 ml) is heated to $100°$ for 30 min. Sodium hydroxide (30%, Note 1) is added until the solution is feebly acidic, and then sodium carbonate is added to pH 7-8. The mixture is extracted with ether, the ether is evaporated, and the residue is sublimed under vacuum to give the product, mp $184°$ (0.26 g, 60% yield) (Note 2).

Notes

1 The oxygen-18 concentration of the water is the same as that used in the rearrangement.

2 Isotopic analysis shows that 87-99% of the oxygen in the product is derived from the water, and not from the sulfuric acid. Various isotopic combinations have been used to prove the origin of the oxygen in the phenol. When hydrochloric acid is used, rather than sulfuric acid, 4-chloroaniline is formed (29% yield) as well as the phenol (43% yield). The intermolecular nature of the reaction has thus been confirmed. Isotopic exchange between starting material and product does not take place under conditions of the rearrangment reaction.

References

[1] I. I. Kukhtenko, "Study of the Mechanism of the Rearrangement of N-Phenylhydroxylamine into p-Aminophenol," *Zh. Org. Khim.*, 7, 330-333 (1971) (English translation pp. 324-326).

3,4-(METHYLENEDIOXY)BENZYL-α-^{13}C ALCOHOL

Procedure[1]

Lithium aluminum hydride (1.7 g, 44 mmol) is added to a solution of 3,4-(methylenedioxy)benzoic-^{13}C acid (5.1 g, 31 mmol) in tetrahydrofuran (45 ml) with stirring in an ice bath. Stirring is continued for 30 min and then at room temperature for 1 hr. The mixture in an ice bath is treated with sulfuric acid (6 *N*) with stirring and extracted with ether. The extract is washed with sodium bicarbonate solution and sodium chloride solution, dried over sodium sulfate, and evaporated to give a white solid (4.5 g, 97% yield).

Other Preparations

By a similar procedure 4-methoxybenzyl-3,5-^{13}C$_2$ alcohol has been prepared from methyl 4-methoxybenzoate-3,5-^{13}C$_2$ and further converted with hydrogen bromide in benzene solution to 1-(bromomethyl)-4-methoxybenzene-3,5-^{13}C$_2$.[2]

References

[1] T. Nagasaki, Y. Katsuyama, and H. Minato, "Synthesis of Radioactive and Stable Isotope-Labelled 1-Ethyl-6,7-methylenedioxy-4(1H)-oxocinnoline-3-carboxylic Acids (Cinoxacin)," *J. Labelled Compds.*, 12, 409-427 (1976).

[2] V. Viswanatha and V. J. Hruby, "Synthesis of [3′,5′-^{13}C$_2$] Tyrosine and Its Use in the Synthesis of Specifically Labeled Tyrosine Analogues of Oxytocin and Arginine-Vasopressin and Their 2-D-Tyrosine Diastereoisomers, *J. Org. Chem.*, 44, 2892-2896 (1979).

3,4-BIS(BENZYLOXY)BENZYL-α-^{13}C ALCOHOL

Procedure[1]

3,4-Bis(benzyloxy)benzoic-^{13}C acid (3.0 g, 9.0 mmol) in tetrahydrofuran (90 ml) is added dropwise to an ice-cold stirred solution of lithium aluminum hydride (1.8 g, 48 mmol) in tetrahydrofuran (175 ml). The mixture is stirred at room temperature for 18 hr, cooled, treated with water (1.9 ml), sodium hydroxide (15%, 1.9 ml), water (5.2 ml), stirred for 15 min, and filtered. The solids are digested with boiling ether, and the organic filtrates are dried over magnesium sulfate and evaporated to give the product as a white solid, mp 66-68° (3.65 g, 95% yield).

References

[1]M. W. Ames and N. Castagnoli, Jr., "The Synthesis of ^{13}C-Enriched α-Methyldopa," *J. Labelled Compds.*, **10**, 195-205 (1974).

α-(METHYL-^{13}C)-3,4-(METHYLENEDIOXY)BENZYL-α-^{13}C ALCOHOL

Procedure[1]

To the Grignard reagent prepared from methyl-^{13}C iodide (5.0 g, 35 mmol) and magnesium (1.0 g, 42 mmol) in ether (60 ml) is added a solution of 3,4-(methylenedioxy)benzaldehyde-^{13}C (3.9 g, 26 mmol) in ether (30 ml) at 0-10° with stirring. The mixture is stirred 1 hr at room temperature, treated dropwise with ammonium chloride (20 g) in water (50 ml), stirred for 15 min, and extracted with ether. The ether extracts are dried over sodium sulfate and evaporated to give the product as a colorless oil (4.3 g, 99% yield).

References

[1]T. Nagasaki, Y. Katsuyama, and H. Minato, "Syntheses of Radioactive and Stable Isotope-Labeled 1-Ethyl-6,7-methylenedioxy-4(1H)-oxocinnoline-3-carboxylic Acids (Cinoxacin)," *J. Labelled Compds.*, **12**, 409-427 (1976).

2-NAPHTHOL-8-^{13}C

Procedure[1]

A solution of 3,4-dihydro-7-methoxy-1(2*H*)-naphthalenone-1-^{13}C (1.20 g, 6.8 mmol) in ether (75 ml) is added dropwise to a stirred, refluxing solution of lithium aluminum hydride (100 mg) in ether. After 1 hr ethyl acetate (3 ml) is added, and the mixture is acidified with hydrochloric acid (2 *N*) and extracted with ether to afford 1,2,3,4-tetrahydro-7-methoxy-1-naphthol-1-^{13}C as an oil (1.21 g, 100% yield) (Note 1). The alcohol is stirred with freshly fused potassium bisulfate (900 mg) at 180° for 1 hr. The ether extracts of an aqueous solution (50 ml) of the cooled reaction mixture are dried over magnesium sulfate and evaporated to give 5,6-dihydro-2-methoxynaphthalene-8-^{13}C (Note 1). The oil (1.0 g) is stirred with palladium-on-charcoal (10%, 1.0 g) at 220° and 550 mm Hg for 1 hr. Ether extraction of the cooled mixture affords crude 2-methoxynaphthalene-8-^{13}C (713 mg, 72% yield, Note 2), which is stirred with pyridine hydrochloride (3 g) at 210° for 6 hr. The cooled reaction mixture is dissolved in hydrochloric acid (1 *N*) and extracted with ether to give the product (669 mg), which is further purified through alkali extraction (613 mg, 95% yield).

Notes

1 The product is found to be pure by GC analysis.
2 The crystals are shown by GC to contain 3% naphthalene-1-^{13}C and 3% unchanged starting material.

References

[1]G. H. Walker and D. E. Hathway, "Synthesis of N-Phenyl-2-[1,4,5,8-^{14}C]naphthylamine, N-Phenyl-2-[8-^{13}C]naphthylamine, and N-[U-^{14}C]Phenyl-2-naphthylamine," *J. Labelled Compds.*, **12**, 199-206 (1976).

1,12-DODECANEDIOL-1,12-^{13}C$_2$

$$HOO*C(CH_2)_{10}*COOH \xrightarrow[\text{THF}]{BH_3 \cdot Me_2S} HO*CH_2(CH_2)_{10}*CH_2OH$$

Procedure[1]

A solution of borane-dimethyl sulfide in tetrahydrofuran (1 M, 40 ml, 40 mmol) is added dropwise under nitrogen at 25° to a stirred solution of dodecanedoic-1,12-^{13}C$_2$ acid (3.2 g, 14 mmol) in tetrahydrofuran (30 ml) during 15 min. The mixture is stirred 8 hr at room temperature, methanol (200 ml) is added slowly, and the solution is evaporated to give the diol as a solid, which is recrystallized from ethanol as needles, mp 81-83° (2.53 g, 90% yield).

References

[1]G. J. Shaw and G. W. A. Milne, "Synthesis and Properties of Dodecane-1,12-^{13}C$_2$ and Dode-cane-1,1,1,12,12,12-^{2}H$_6$," J. Labelled Compds., 12, 557-563 (1976).

2-PHENYL-3-(PHENYL-1-^{13}C)OXIRANE-2-^{13}C
(^{13}C$_2$-STILBENE OXIDE)

Procedure[1]

1-Phenyl-2-(phenyl-1-^{13}C)ethene-1-^{13}C (23.0 mg, 0.127 mmol) and 3-chloroperoxy-benzoic acid (24.1 mg, 0.140 mmol) (Note 1) in benzene (6 ml) are stirred at 35° for 72 hr. The solvent is evaporated to leave the white, crystalline, crude product (mp 69-78°), which is purified by preparative thin layer chromatography (silica gel GF254, carbon tetrachloride); mp 76-77° (12.2 mg, 44% yield).

Notes

1   The peroxy acid is previously washed with phosphate buffer (pH 7.5) to remove 3-chlorobenzoic acid.

References

[1]G. A. Braden and U. Hollstein, "Synthesis of 1-Phenyl-2-phenyl-1-^{13}C-ethene-1-^{13}C (trans-Stilbene) and Derivatives," J. Labelled Compds., 12, 507-516 (1976).

7-BENZO[a]PYRENOL-7-[13]C

Procedure[1]

A mixture of 9,10-dihydrobenzo[a]pyren-7(8H)-one-7-[13]C (701 mg, 2.60 mmol), palladium-on-carbon (10%, 150 mg), and methylnaphthalene (10 ml) is heated at reflux for 18 hr, cooled, diluted with boiling xylene (15 ml), and filtered with the aid of additional hot xylene (15 ml). The solution is concentrated to near dryness, and the solid is recrystallized from xylene to give the product as fine, yellow needles, mp 216-218° (dec.) (350 mg, 50% yield).

References

[1]W. P. Duncan, W. C. Perry, and J. F. Engel, "Labeled Metabolites of Polycyclic Aromatic Hydrocarbons IV. 7-Hydroxybenzo[a]pyrene-7-^{13}C," *J. Labelled Compds.*, **12**, 275-280 (1976).

6-BENZO[a]PYRENEMETHANOL-α-[13]C

Procedure[1]

6-Benzo[a]pyrenemethanol-α-[13]C 6-Benzo[a]pyrenecarboxaldehyde-[13]C (3.25 g, 11.6 mmol), aluminum isopropoxide (16.3 g, 79 mmol), and 2-propanol (325 ml) are distilled slowly with a Hahn condenser until no more acetone is collected. The remaining solution (150 ml) is poured into cold hydrochloric acid (5%, 600 ml).

The crude product is recrystallized three times from benzene to give the product, mp 228-229° (dec.) (Note 1) (3.1 g, 95% yield).

6-(Chloromethyl-^{13}C)benzo[a]pyrene The above-mentioned alcohol (0.20 g, 0.71 mmol), thionyl chloride (0.20 g, 1.7 mmol), and benzene (7 ml) are refluxed for 30 min with stirring. The solvent is removed, and the crude material is recrystallized twice from benzene to give the product, mp 218-219° (0.17 g, 80% yield).

Notes

1 In a sealed tube, the mp is 230-231° without decomposition, followed by solidification and remelting at 290° with decomposition.

References

[1]R. E. Royer, G. H. Daub, D. L. Vander Jagt, "Synthesis of Carbon-13 Labelled 6-Substituted Benzo[a]pyrenes," *J. Labelled Compds.*, **12**, 377-380 (1976).

Amines and Hydrocarbons

Labeling of amines at the functional group, with either carbon or nitrogen, most frequently starts with the amides, which may be either reduced or rearranged. The former is more common for carbon and the latter for nitrogen. Synthesis of saturated hydrocarbons usually proceeds by reduction of carbonyl compounds or carboxyl derivatives, while unsaturated hydrocarbons are often obtained by a Wittig-type synthesis or dehydration of an alcohol from a Grignard reaction. Amines and hydrocarbons are generally considered to be structurally "simpler" than compounds with functional groups at higher oxidation levels; their synthesis, however, is often not so simple, owing to their dependence on availability of a "more complicated" labeled compound as a starting material. As is so often the case with isotopic syntheses, preparation of a particular material depends largely on the availability of a particular carbonyl or carboxyl compound as a starting material. The development of schemes for labeling at particular molecular locations or at multiple sites can involve a wide variety of reactions, as exemplified in methods applied for introduction of several carbon-13 atoms into alkanes using a minimum of different chain prolongation reactions.[1]

References

[1] A. Heusler, P. Ganz, and T. Gaumann, "Synthesis of n-Alkane Derivatives Labelled with Several ^{13}C," J. Labelled Compds., 11, 37-42 (1975).

METHANE-^{13}C

$$*CO \; or \; *CO_2 \quad \xrightarrow[\text{300-400}^\circ]{\text{H}_2, \text{Ru-Al}_2\text{O}_3} \quad *CH_4$$

Procedure[1]

Carbon-^{13}C dioxide (94.3 g, 2.10 mol, 95 psig) (Note 1) is introduced into an evacuated 7-liter converter (Note 2) and the catalyst is heated to approx. 300° (Note 3). The lower portion of the converter is cooled in ice, and hydrogen is added from a calibrated 8-liter cylinder (45 psi/mol) in increments of 20 to 50 psi, which cause a temperature rise of the catalyst bed to 350-400° (Note 4). After addition of the hydrogen (8.5 mol, 380 psi decrease in cylinder pressure) over approx. 5 hr (Note 5), the bottom of the reactor is cooled, first with Dry Ice and then with liquid nitrogen, and the excess hydrogen is pumped out through a liquid-nitrogen-cooled charcoal trap (Note 6). The cooling baths are removed from the converter and from the charcoal trap, and the product is transferred into a liquid-nitrogen-cooled 1-liter gas cylinder (34.1 g, 725 psig, 96% yield) (Note 7).

Notes

1 Carbon-^{13}C monoxide can also be used.
2 The converter[2] relies on convective circulation of the gases through the electrically heated catalyst (0.5% ruthenium on alumina, 1/8-in. pellets) contained in stainless steel screen located on the central axis of the 7-liter, vertically oriented, cylindrical vessel. A similar, but smaller, reactor of approx. 1-liter capacity has also been used[1] and is described in Chapter 4, Methanol-^{13}C, Method I.
3 The pressure rises to approx. 130 psig.
4 Subsequent additions of hydrogen are made after the catalyst has returned to approx. 300°.
5 Temperature rise on addition of hydrogen no longer occurs.
6 The vapor pressure of methane at liquid nitrogen temperature is quite high (approx. 10 mm Hg); therefore, pumping is not prolonged or the capacity of the charcoal may be exceeded.
7 Residual methane remaining in the apparatus amounts to only a fraction of a percent. Infrared analysis of the product shows no carbon monoxide or dioxide. Gas chromatography indicates a small amount of hydrogen.

Other Preparations

By the same procedure have been prepared methane-^{13}C-d_4 and methane-d_4;[1] carbon monoxide as starting material requires less deuterium.

References

[1] D. G. Ott, V. N. Kerr, T. G. Sanchez, and T. W. Whaley, "Syntheses with Stable Isotopes: Sodium Cyanide-^{13}C, Methane-^{13}C, Methane-^{13}C-d_4, and Methane-d_4," *J. Labelled Compds.*, 17, 255-262 (1980).
[2] R. W. Buddemier, A. Y. Young, A. W. Fairhall, and J. A. Young, "Improved System of Methane Synthesis for Radiocarbon Dating," *Rev. Sci. Instrum.*, 41, 652-565 (1970).

IODOMETHANE-^{13}C

$$*CH_3OH \xrightarrow[H_2O]{HI} *CH_3I$$

Procedure[1]

Methanol-^{13}C (7.47 g of solution containing 11.9% water, 0.20 mol) and cold hydriodic acid (57%, 132 ml, 1 mol) in a 100-ml 3-neck flask fitted with an argon inlet, thermometer, and spiral reflux condenser (air-cooled) that leads to a Dry Ice-cooled trap (100-ml round-bottom flask) followed by a smaller second trap and silicone oil bubbler are slowly heated in an oil bath over 4 hr to 70°. Temperature of the bath is raised to 85° over the next several hours (Note 1). Argon flushing (approx. 5 cc/min) is begun and maintained for an additional hour. The first cold trap, containing product (25.52 g), is replaced with another, and gentle argon flushing is continued overnight with the reaction flask at ambient temperature. An additional amount of product (2.48 g) (Note 2) is combined with the main fraction, treated with molecular sieve (Types 3A and 5A, 10 g each), and subjected to bulb-to-bulb vacuum transfer to give the clear, colorless product (26.95 g, 94% yield) (Note 3).

Notes

1 The temperature of the reaction mixture remains at approx. 35-40° for the first 6 hr (some refluxing is observed), then rises to approx. 70°.

2 No material is collected in the second cold trap.

3 The product is pure by ^{1}H-NMR analysis, and is stored over copper shot. Precautions with respect to the volatility of the material must be observed during transfers.

Other Preparations

By essentially the same procedure iodomethane-^{13}C has been prepared in yields of 79%[2], 73%[3], 62%[4], and 93%[5]. The variation in yields is probably a consequence of the volatility of the product. The reaction should be conducted in a manner such that unreacted methanol is not carried out of the reaction flask during the reaction, and that the product is efficiently condensed. Preparation using the phosphorus-iodine method[6] has also been used,[7] but the apparatus and procedure are more complex.

References

[1]D. G. Ott, Los Alamos Scientific Laboratory, unpublished results (1979).

[2]T. W. Whaley, G. H. Daub, V. N. Kerr, T. A. Lyle, and E. S. Olson, "Syntheses with Stable Isotopes: DL-Valine-^{13}C$_3$," *J. Labelled Compds.*, **16**, 809-817 (1979).

[3]M. Pomerantz and R. Fink, "Synthesis of Diethyl 1-[(^{13}C)Methyl]-3-phenyl(1,3-^{13}C$_2$)bicyclo-[1.1.0]butene-exo, exo-2,4-dicarboxylate," *J. Labelled Compds.*, **16**, 275-286 (1979).

[4] M. A. G. El-Fayoumy, H. C. Dorn, and M. A. Ogliaruso, "Preparation of Trimethylsulfonium-^{13}C Iodide," *J. Labelled Compds.*, **13**, 433-436 (1977).

[5] J. D. Roberts, R. E. McMahon, and J. S. Hine, "Rearrangements of Carbon Atoms in *t*-Butyl and *t*-Amyl Derivatives," *J. Am. Chem. Soc.*, **72**, 4237-4244 (1950).

[6] A. Murray and D. L. Williams, *Organic Syntheses with Isotopes*, Interscience, New York, 1958, pp. 862-863.

[7] T. W. Whaley and D. G. Ott, in "Annual Report of the Biological and Medical Research Group (H-4)," C. R. Richmond and G. L. Voelz, Eds., Los Alamos Scientific Laboratory Report LA-4923-PR, April 1972, p. 112.

ACETYLENE-$^{13}C_2$

$$2\text{*CO}_2 \xrightarrow[\text{2. H}_2\text{O}]{\text{1. Li, 650-760}°} \text{H*C}\equiv\text{*CH}$$

$$\xrightarrow[\text{Dioxane}]{\text{H}_2\text{NCH}_2\text{CH}_2\text{NHLi}} \text{Li*C}\equiv\text{*CH}\cdot\text{H}_2\text{NCH}_2\text{CH}_2\text{NH}_2$$

Procedure[1]

Acetylene-$^{13}C_2$ The reaction vessel (Note 1) containing lithium shot (46.5 g, 6.7 mol) is evacuated, heated to 620°, and isolated from the vacuum pump. Carbon-^{13}C dioxide (42.8 g, 0.953 mol) is pulsed into the chamber in increments causing a rise of 5-10 in. Hg over 1.25 hr at a rate such that the temperature remains between 640° and 670°. The reaction chamber is opened to the vacuum pump and heated to 760° for 1.25 hr. The reaction is cooled and then hydrolyzed by the cautious addition of water (approx. 2 liter). The mixture of acetylene, hydrogen, and water vapor that is liberated is passed through a Dry Ice trap, a U-tube containing molecular sieve (Type 3A), anhydrous lithium hydroxide (Lithasorb), and two traps cooled in liquid nitrogen (Note 2). The product, which collects in the traps, is sublimed into two 4.5-liter storage vessels (0.447 mol, 94% yield) (Note 3).

Lithium acetylide-$^{13}C_2$ ethylenediamine complex A slurry of *N*-lithioethylenediamine (26.4 g, 0.399 mol) (Notes 4 and 5) in dioxane (200 ml) at 15° under an argon atmosphere in a 500-ml 3-neck flask fitted with a magnetic stirrer, thermometer, condenser with a vacuum outlet and balloon, and a gas inlet tube is partially evacuated. Acetylene-$^{13}C_2$ (0.210 mol) from a storage vessel is transferred to a liquid-nitrogen-cooled trap and then allowed to sublime into the ice-cooled reaction flask at a rate such that the temperature remains at 15°, and a slight positive pressure (indicated by the balloon) is maintained. The mixture is seeded with a few crystals of the lithium acetylide ethylenediamine complex, and a second portion of acetylene (0.209 mol) is transferred to the reaction as before. After the acetylene is consumed the mixture, containing a tan precipitate, is stirred at room temperature for 3 hr, poured into petroleum ether (200 ml) (Note 5), filtered, and washed with petroleum ether to give the product as a light-tan powder (32.4 g, 83% yield) (Note 6).

Notes

1   The stainless steel reaction chamber is from the Alford Instrument Co., Knoxville, Tennessee, and is heated by two 1-kilowatt semicylindrical heating elements. A diagram of the apparatus, traps, and storage vessels is given.[1]
2   Hydrolysis and collection of the majority of the product requires approx. 8 hr; after standing overnight the remainder of the acetylene is collected.
3   The yield is calculated from the pressure. ^{1}H-NMR data are given.
4   The air-sensitive lithium compound is obtained from freshly prepared lithium dispersion and ethylenediamine in benzene.
5   Isolation of the product is conducted in an argon-flushed dry box.
6   The complex is stable to dry air; hydrolysis liberates acetylene-$^{13}C_2$ in 95% yield.

References

[1]T. W. Whaley and D. G. Ott, "Syntheses with Stable Isotopes: Acetylene-$^{13}C_2$ and Lithium Acetylide-$^{13}C_2$ Ethylenediamine Complex," *J. Labelled Compds.*, **10**, 461-468 (1974).

1,3-DIBROMOPROPANE-1-^{13}C

$$\text{EtOOCCH}_2\text{*COOEt} \xrightarrow[\text{2. HBr, H}_2\text{SO}_4]{\text{1. LiAlH}_4, \text{Et}_2\text{O}} \text{BrCH}_2\text{CH}_2\text{*CH}_2\text{Br}$$

Procedure[1]

Diethyl malonate-1-^{13}C (19.4 g, 0.121 mol) in ether (75 ml) is added to a stirred solution of lithium aluminum hydride in ether (1.25 M, 200 ml) at 0° over a period of 45 min. The mixture is refluxed for 2 hr, and water (10 ml) is added dropwise over 10 min, followed by sulfuric acid (conc., 26.6 ml, 0.5 mol). The ether is distilled at atmospheric pressure, hydrobromic acid (63%, 114 ml, 1.5 mol) (Note 1) and sulfuric acid (conc., 110 ml) are added, and the mixture is refluxed for 6 hr at 115°. The lower phase of the distillate from distillation of the reaction mixture at atmospheric pressure (Note 2) is separated, washed with water, sulfuric acid, and sodium carbonate (10%), dried over magnesium sulfate, and distilled to give the product, bp 81-88° at approx. 46 mm Hg (14.2 g, 58% yield (Note 3).

Notes

1   Concentration of the hydrobromic acid in the reaction mixture is 48%.
2   Distillation is continued until oily drops no longer are present in the distillate.
3   The product is pure by ^{1}H-NMR and IR.

References

[1]B. Bak and J. J. Led, "Preparation of [1-^{13}C], [3-^{13}C], [1-D] and [3-D] Enriched Cyclobutenes," *J. Labelled Compds.*, **4**, 22-27 (1968).

^{13}C-α-METHYLSTYRENE

Method I

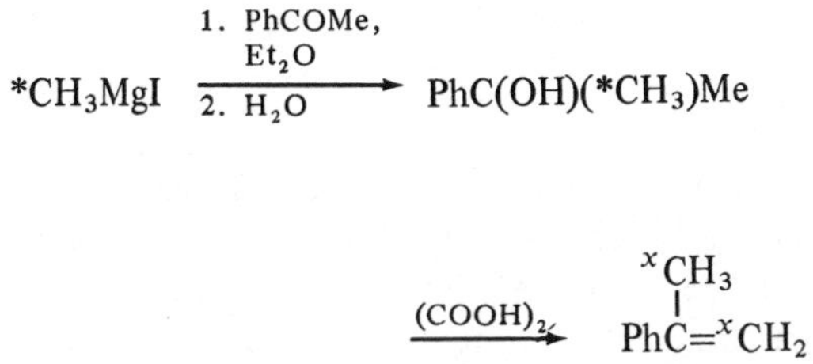
1. PhCOMe,
Et$_2$O
*CH$_3$MgI
2. H$_2$O
PhC(OH)(*CH$_3$)Me
(COOH)$_2$
*CH$_3$
PhC=*CH$_2$

Procedure[1]

2-Phenylpropene-1,3-$^{13}C_1$ A solution of acetophenone (24 g, 0.20 mol) in ether (30 ml) is added dropwise with stirring to the Grignard reagent prepared from methyl-^{13}C iodide (28 g, 0.20 mol) (Note 1). The mixture is warmed on a water bath for 2 hr, hydrolyzed with ice (20 g), and extracted several times with ether. The extract is dried over sodium sulfate and fractionated at reduced pressure to give 2-phenyl-2-propanol-1-^{13}C, bp 89° at 13 mm Hg (24 g, 75% crude yield) (Note 2). A mixture of the alcohol and oxalic acid (2.5 g) is slowly heated to reflux in an oil bath at 150°, cooled, and treated with ether (100 ml). The aqueous phase is separated, and the ether layer is washed with water (3 × 20 ml), dried over sodium sulfate, and distilled through a column to afford the product, bp 140° (16.7 g, 71% yield).

Method II

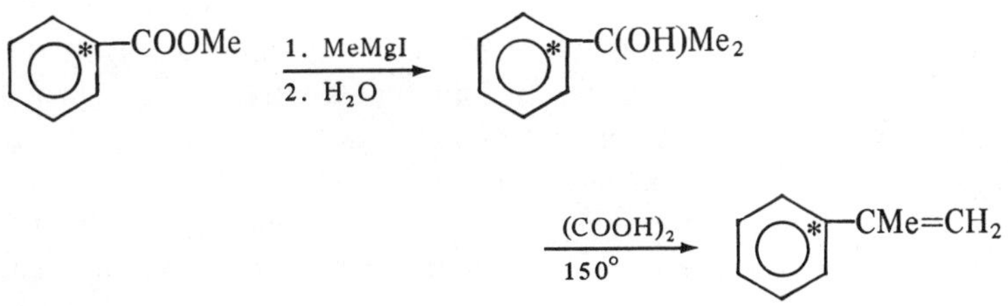
COOMe
1. MeMgI
2. H$_2$O
C(OH)Me$_2$
(COOH)$_2$
150°
CMe=CH$_2$

Procedure[1]

2-(Phenyl-1-^{13}C)propene To the Grignard reagent prepared from methyl iodide (9.3 g, 65 mmol) and magnesium (4.0 g, 0.17 mol) in ether (40 ml) is added dropwise a solution of methyl benzoate-1-^{13}C (3.5 g, 25 mmol) in ether (40 ml). The mixture is refluxed for 2 hr and then treated slowly with ice water until the ether no longer boils. The ether solution is decanted, dried over sodium sulfate, and evaporated. The residue of α,α-dimethylbenzyl-1-^{13}C alcohol is dehydrated with oxalic acid as described in Method I to afford the product (2 g, 63% yield).

Notes

1 A white solid appears on warming.
2 Further purification to remove contaminating acetophenone is not carried out since it is removed more simply in the next step.

Other Preparations

Styrene-α-^{13}C, styrene-β-^{13}C, and styrene-α,β-^{13}C$_2$ have been prepared by dehydrohalogenation of the corresponding ^{13}C-1-bromo-2-phenylethanes with potassium *t*-butoxide in *t*-butanol at 75°.[2] The intermediate ^{13}C-1-phenylethanols are prepared from benzaldehyde-^{13}C and methyl-^{13}C iodide.

References

[1] B. Stuetzel, W. Ritter, and K-F. Elgert, "^{13}C-Enriched Monomers for NMR-Spectroscopy of Polymers," *Angew. Makromol. Chemie*, **50**, 21-41 (1976).
[2] A. Venema, N. M. M. Nibbering, and T. J. DeBoer, "Mass Spectrometry of Aralkyl Compounds with a Functional Group–X," *Org. Mass Spectrom.*, 3, 1589-1592 (1970).

1-PHENYLPROPYNE-^{13}C$_3$

$$Ph*CH_2*CO*CH_3 \xrightarrow[\substack{2.\ Hg_2(COCF_3)_2, \\ Et_2O}]{1.\ H_2NNH_2,\ 100°} \quad \substack{Ph*C\equiv *C*CH_3 \\ + \\ Ph*CH=*C=*CH_2} \xrightarrow[]{\substack{KOBu\text{-}t, \\ t\text{-}BuOH, \\ DMSO}}$$

Procedure[1]

1-Phenyl-2-propanone-^{13}C$_3$ (4.0 g, 29 mmol) and hydrazine (95%, 4.0 g, 0.13 mol) are refluxed at 100° for 3 hr. The reaction mixture is extracted with ether, and the extract is washed with a small amount of water, dried over potassium carbonate, and evaporated to give the hydrazone (4.0 g, 93% yield).

Into a 500-ml 3-neck flask equipped with stirrer, condenser, drying tube, and addition funnel are placed mercurous trifluoroacetate (34 g, 54 mmol) and ether (100 ml). The above-mentioned hydrazone (4.0 g, 27 mmol) in ether (100 ml) is added with stirring under reflux over 1 hr, and stirring is continued for 0.5 hr. The ether layer is separated, washed with ammonium hydroxide (5%) and water, dried over potassium carbonate, and evaporated to give a red oil (2.5 g), which is chromatographed on alumina (0.75-in. diameter column, 40 g) using pentane. Evaporation of the first 500 ml of eluant gives a mixture (1.0 g) of 1-phenylpropyne-^{13}C$_3$ and 1-phenylpropadiene-^{13}C$_3$, which is stirred for 3 hr at room temperature with potassium *t*-butoxide (50 mg), *t*-butanol (1.4 g), and dimethyl sulfoxide (2 ml). Water (50 ml) is added, and the mixture is extracted with pentane (5 × 30 ml). The extract is washed with water, dried over magnesium sulfate, and evaporated to give the product (600 mg, 18% yield).

Other Preparations

The above product is converted in 21% yield to diethyl 1-(methyl-^{13}C)-3-phenylbicyclo[1.1.0] butane-exo,exo-2,4-dicarboxylate-1,3-^{13}C$_2$.

References

[1]M. Pomerantz and R. Fink, "Synthesis of Diethyl 1-[(^{13}C)Methyl]-3-phenyl(1,3-^{13}C$_2$)bicyclo-[1.1.0] butane-exo,exo-2,4-dicarboxylate," *J. Labelled Compds.*, **16**, 275-286 (1979).

^{13}C-1,3-BUTADIENE

$$\text{EtOO*CCH}_2{}^\bullet\text{CH}_2\text{*COOEt} \xrightarrow[\text{Et}_2\text{O}]{\text{LiAlH}_4} \text{HO*CH}_2{}^\bullet\text{CH}_2\text{CH}_2\text{*CH}_2\text{OH}$$

$$\xrightarrow[\text{P}_2\text{O}_5]{\text{KI, H}_3\text{PO}_4} \text{I*CH}_2{}^\bullet\text{CH}_2\text{CH}_2\text{*CH}_2\text{I} \xrightarrow[\substack{2.\ \text{Ag}_2\text{O},\\ \text{H}_2\text{O}}]{1.\ \text{Me}_3\text{N}}$$

$$\overset{+}{\text{Me}_3\text{N}}\text{*CH}_2{}^\bullet\text{CH}_2\text{CH}_2\text{*CH}_2\overset{+}{\text{NMe}_3} + 2\text{OH}^- \xrightarrow{100\text{-}250^\circ} \text{*CH}_2={}^\bullet\text{CHCH}=\text{*CH}_2$$

Procedure[1]

1,4-Diiodobutane-2-^{13}C or 1,4-diiodobutane-1,4-^{13}C$_2$ Diethyl succinate-2-^{13}C or diethyl succinate-1,4-^{13}C$_2$ (36.5 g, 0.21 mol) in ether (200 ml) is added dropwise to a vigorously stirred suspension of lithium aluminum hydride (40 g, 1.05 mol) in ether (2 liter). The mixture is refluxed 2 hr, excess hydride is decomposed by dropwise addition of water and sulfuric acid (15%, 1.5 liter), and the 1,4-butanediol-2-^{13}C or 1,4-butanediol-1,4-^{13}C$_2$ is isolated by continuous extraction with ether. With stirring and exclusion of moisture finely powdered potassium iodide (160 g, 0.96 mol) and the crude diol are added to a mixture of phosphoric acid (85%, 124 g) and phosphorus pentoxide (29 g). The mixture is heated under reflux for 5 hr, cooled, and treated with water (60 ml) and ether (100 ml). The ether layer is decolorized with sodium thiosulfate, dried over sodium sulfate, and distilled, bp 120-124° at 14 mm Hg (46.6 g, 67% yield based on the ester).

1,3-Butadiene-2-^{13}C or -1,4-^{13}C$_2$ The diiodobutane (46.6 g, 0.15 mol) and trimethylamine-saturated ethanol (500 ml) are refluxed 2 hr with exclusion of moisture. The mixture, containing precipitated product, is treated with dry ether (2 liter), kept at 0° for 5 hr, and filtered. The filtrate is evaporated at reduced pressure, and the residual unreacted diiodobutane is again treated with trimethylamine. The combined products after filtration are washed with ether and dried under vacuum to give tetramethylene-2-^{13}C-bis(trimethylammonium) or tetramethylene-1,4-^{13}C$_2$-bis(trimethylammonium) diiodide (59.5 g, 93% yield). The salt is converted to tetramethylene-2-^{13}C-bis(trimethylammonium) dihydroxide or tetramethylene-1,4-^{13}C$_2$-bis(trimethylammonium) dihydroxide by stirring for

16 hr in water (450 ml) with the washed silver oxide obtained from silver nitrate (90 g, 0.53 mol) and sodium hydroxide (8%, 450 ml). The filtered solution is heated in a bath while the temperature is gradually raised to 250°. A slow stream of nitrogen is used, the water vapor is condensed, and the trimethylamine is separated by passage through sulfuric acid (3 N). Finally, the product is dried with calcium chloride and condensed in a Dry Ice-methanol trap (6.7 g, 90% yield) (Note 1).

Notes

1 The purity, determined by gas chromatography, is 98%.

References

[1]B. Stuetzel, W. Ritter, and K-F. Elgert, "^{13}C-Enriched Monomers for NMR-Spectroscopy of Polymers," *Angew, Makromol. Chem.*, **50**, 21-41 (1976).

CYCLOBUTENE-1,3-^{13}C$_1$

Procedure[1]

1,3-Dibromopropane-1-^{13}C (14.2 g, 70 mmol) is converted to diethyl 1,1-cyclo-butanedicarboxylate-2-^{13}C (10.1 g, 50.5 mmol), which is hydrolyzed to 1,1-cyclo-butanedicarboxylic-2-^{13}C acid (6.54 g, 45.4 mmol) and decarboxylated to cyclobutanecarboxylic-2-^{13}C acid (4.44 g, 44 mmol). Cyclobutanamine-2-^{13}C hy-drochloride (3.98 g, 37.0 mmol) is obtained from the acid and further converted to (cyclobutyl-2-^{13}C)trimethylammonium iodide (Note 1). Treatment of the mixture with an excess of silver oxide and water and filtration affords an aqueous solution of the quaternary hydroxide (40 ml), which is added dropwise into an evacuated, 2-neck, 100-ml flask at 145°. The pressure is kept at 1-5 mm Hg by pumping through a trap containing a diethylene glycol monobutyl ether solution of sulfuric acid (1 N, 100 ml). After the pyrolysis, the closed trap is slowly heated to 5° and then cooled to −40° (Note 2). The cyclobutenes, accompanied by a trace of ethyl-ene, are distilled (bp 2°), cooled to −100°, and fractionated for removal of ethylene to afford the product (1.24 g, 33% yield) (Note 3).

Notes

1 The iodide (approx. 8 g) is accompanied by potassium iodide and chloride (approx. 4 g).

2 The trap solution dissolves the water and cyclobutene while binding trimethylamine chemically.

3 The product is pure by IR analysis. ^{1}H-NMR analysis shows almost equal amounts of the two isotope isomers.

References

[1] B. Bak and J. J. Led, "Preparation of [1-^{13}C], [3-^{13}C], [1-D] and [3-D] Enriched Cyclobutenes," *J. Labelled Compds.*, 4, 22-27 (1968).

CYCLOPENTADIENYLTHALLIUM-$^{13}C_1$

Procedure[1]

1,2-Dibromocyclopentane-1-^{13}C Cyclopentanol-1-^{13}C (9.1 g, 0.10 mol), phosphoric acid (85%, 40 g), and potassium pyrosulfate (24 g) are heated slowly with magnetic stirring to 85-90° and then, after initial frothing, to 140°. The evolved cyclopentene-1-^{13}C (bp 37-40°, 6.2 g, 86% yield) after passage through a small condenser is condensed in a Dry Ice trap, dried over calcium sulfate, dissolved in carbon tetrachloride (100 ml), chilled in Dry Ice, and treated slowly with a solution of bromine (16 g, 0.1 mol) in carbon tetrachloride (100 ml). The excess bromine and solvent are removed at reduced pressure, and the residue is distilled to give the product, bp 49-51° at 0.3 mm Hg (17.9 g, 75% yield from cyclopentanol).

Cyclopentadienylthallium-$^{13}C_1$ 1,2-Dibromocyclopentane-1-^{13}C (6.7 g, 25 mmol) is introduced from a syringe through a serum stopper into a rapidly stirred solution of freshly distilled quinoline (15 ml) in hexamethylphosphoric triamide (20 ml) at 185-190°. Ethanol is added intermittently from a second syringe to flush the apparatus of product, which is condensed in an ice-cooled trap. The cyclopentadiene-$^{13}C_1$ is added to an aqueous solution of potassium hydroxide (10 g, 0.16 mol) and thallous acetate (7.1 g, 27 mmol), which is stoppered, shaken vigorously, and chilled in an ice bath. The precipitate is filtered, washed with water and cold methanol, air dried, and sublimed (85-100° at 0.01 mm Hg) to afford the product (2.1 g, 31% yield).

Other Preparations

Cyclopentadiene-vic-$^{13}C_2$ has been prepared by pyrolysis of 2-acetoxybicyclo-[2.2.1]heptane-2,3-$^{13}C_2$, which is obtained from catalytic hydrogenation of the product from condensation of cyclopentadiene and vinyl-$^{13}C_2$ acetate. The latter compound is prepared by mercuric acetate catalyzed addition of acetic acid to acetylene-$^{13}C_2$. Photoisomerization of the cyclopentadiene-vic-$^{13}C_2$ forms a mixture of $^{13}C_2$-cyclopentadiene (non-vic), $^{13}C_2$-bicyclo[2.1.0]pent-2-ene, and $^{13}C_2$-tricyclo[2.1.0.0^{2,5}]pentane.[2]

References

[1]S. D. Larsen, P. J. Vergamini, and T. W. Whaley, "Synthesis of Cyclopentadienyl-x-^{13}C Thallium," J. Labelled Compds., 11, 325-332 (1975).
[2]G. D. Andrews and J. E. Baldwin, "Phototransposition of Carbon Atoms in Cyclopentadiene. Photoisomerization of Cyclopentadiene to Tricyclo[2.1.0.0^{2,5}]pentane," J. Am. Chem. Soc., 99, 4851-4853 (1977).

$^{13}C_1$-1-METHYLCYCLOHEXENE

Method I

Procedure[1]

1-Methylcyclohexene-1-^{13}C To 1-methylcyclohexanol-1-^{13}C (12.7 g, 0.11 mol) is added phosphoric acid (85%, 2 ml), and the mixture is heated in a bath at 140°. A mixture of product and water distills. The aqueous phase is separated, and the product is dried over calcium hydride and distilled, bp 102-110° (6.5 g, 61% yield).

Method II

Procedure[2]

1-(Methyl-^{13}C)cyclohexene   A crystal of iodine is added to 1-(methyl-^{13}C)cyclohexanol (Note 1), and the mixture is heated at 140-145° for 2 hr. The products are distilled at 0.5 mm Hg into a liquid-nitrogen-cooled receiver, and the organic layer is dried over sodium sulfate and redistilled, bp 98-101° (1.53 g). The reaction residue is subjected to the procedure a second time to afford additional product (total yield 2.12 g, 69%) (Note 2).

Notes

1 The crude product from reaction of ethyl acetate-2-¹³C (32 mmol) with the Grignard reagent from 1,5-dibromopentane (95 mmol) is used.
2 The yield is based on ethyl acetate-2-¹³C. ¹H-NMR data are given.

Other Preparations

1-Methylcyclohexene-1-¹³C has also been prepared by the procedure of Method II in 59%[2] and 90%[3] yields.

References

[1] B. Stuetzel, W. Ritter, and K-F. Elgert, "¹³C-Enriched Monomers for NMR-Spectroscopy of Polymers," *Angew. Makromol. Chem.,* **50,** 21-41 (1976).
[2] G. A. Braden and U. Hollstein, "Synthesis of 1-Phenyl-2-phenyl-1-¹³C-ethene-1-¹³C (*trans*-Stilbene) and Derivatives," *J. Labelled Compds.,* **13,** 507-516 (1976).
[3] R. J. Dickinson and D. Williams, "Investigation of Possible Isomerization Reactions in Benzene, Iodobenzene, and Phenol upon Electron Impact," *J. Chem. Soc. B,* **1971,** 249-251.

BENZENE-1-¹³C-1-*d*

Procedure[1]

Magnesium turnings (0.1 g, 4 mmol) are stirred under ether (0.5 ml), and methyl iodide (0.1 g, 0.7 mmol) is added. After reaction ceases iodobenzene-1-¹³C (0.07 g, 0.34 mmol) in ether (0.5 ml) is added. After ther reaction is complete all volatile components are evaporated under vacuum. Ether (0.5 ml) and deuterium oxide (1.0 ml, 50 mmol) are added in turn. All volatile products are removed under vacuum and trapped at −196°. The product is isolated by preparative gas chromatography (approx. 70% yield).

References

[1] R. J. Dickinson and D. Williams, "Investigation of Possible Isomerization Reactions in Benzene, Iodobenzene, and Phenol upon Electron Impact," *J. Chem. Soc. B,* **1971,** 249-251.

BENZENE-1,2-$^{13}C_2$-3,4,5,6-d_4

Procedure[1]

Lead tetraacetate (3.4 g, 7.7 mmol) is added to 4-cyclohexene-1,2-dicarboxylic-1,2-$^{13}C_2$-3,3,4,5,6,6-d_6 acid (610 mg, 3.6 mmol) (Note 1) and dimethyl sulfoxide (5 ml) in a 25-ml 2-neck flask equipped with a nitrogen inlet and reflux condenser leading to a Dry Ice trap and bubbler (Note 2). With nitrogen flushing the magnetically stirred mixture is heated slowly in an oil bath to 65° (Note 3), the flow of condenser water is stopped, and heating is continued at 70-75° for 24 hr. The trap is connected to a vacuum line, and the contents are distilled through potassium hydroxide pellets into a tube containing tetrachloro-o-quinone (0.5 g, 2 mmol) (Note 4). The tube is sealed, placed in an oven at 75° for 20 hr, cooled to −20°, and opened. The product is collected on a vacuum manifold, mp 3° (203 mg, 70% yield) (Note 5).

Notes

1 The acid is obtained by hydrolysis of 4-cyclohexene-1,2-dicarboxylic-1-2-$^{13}C_2$-3,3,4,5,6,6-d_6 anhydride in water on a steam bath for 5 hr (90% yield). The anhydride is prepared by Diels-Alder synthesis from the deuterated diene and maleic-2,3-$^{13}C_2$ anhydride (87% yield).

2 Attempted decarboxylation and dehydrogenation with phosphorus pentoxide at 200° gives a yield of only 39% and scrambling of the deuterium positions.

3 Evolution of carbon dioxide is rapid.

4 When the dihydrobenzene does not contain the deuterium, it is rapidly oxidized to benzene with excess lead tetraacetate; however, the deutero analogue oxidizes only very slowly.

5 Additional product (45 mg) of lower purity can be obtained from the reaction mixture of flushing at higher temperature.

Other Preparations

Benzene-$^{13}C_6$ has been prepared in high yield[2,3] by trimerization of acetylene-$^{13}C_2$ in the presence of vanadium oxide-on-alumina catalyst, a method develped for radiocarbon dating.[4]

References

[1]R. N. Renaud and L. C. Leitch, "Synthesis of 3,4,5,6-Tetradeuteriobenzene-1,2-$^{13}C_2$," *J. Labelled Compds.*, 9, 145-148 (1973).

[2]T. W. Whaley and D. G. Ott, in "Annual Report of the Biological and Medical Research Group (H-4)," C. R. Richmond and G. L. Voelz, Eds., Los Alamos Scientific Laboratory Report LA-5227-PR, March 1973, p. 105.

[3]D. G. Ott, "Carbon-13 Labelled Compounds," *Kagaku No Ryoiki Zokan,* **107**, 96 (1975).

[4]J. E. Noakes, S. M. Kim, and J. J. Stipp, "Chemical and Counting Advances in Liquid Scintillation Age Dating," in *Proceedings of the Sixth International Conference on Radiocarbon and Tritium Dating,* Pullman, Washington (1965), p. 68.

NITROBENZENE-1,4-^{13}C$_1$

Procedure[1]

Diazotization[2] of 4-nitroaniline-1,4-^{13}C$_1$ (204 mg, 1.48 mmol) is followed by deamination.[3] Cold hypophosphorous acid is added directly to the diazotization mixture at ice temperature, which is then stirred for 20 min, stored at 2° for approx. 48 hr, and extracted with ether (5 × 50 ml). The extract is washed with sodium hydroxide (20%), dried, and concentrated to afford the product, which is purified by preparative gas chromatography (5% Carbowax on Chromasorb-P, 180°) (96 mg, 53% yield).

References

[1]G. L. Swartz and W. M. Gulick, Jr., "Synthesis of Carbon-13 Enriched Nitrobenzene," *J. Labelled Compds.,* **11**, 525-534 (1975).

[2]A. W. Ruddy and E. B. Starkey, "*p*-Nitrophenylarsonic Acid," *Org. Syn.,* Col. Vol. **3**, 665-667 (1955).

[3]N. Kornblum, "3,3'-Dimethoxybiphenyl and 3,3'-Dimethylbiphenyl," *Org. Syn.,* Col. Vol. **3**, 295-299 (1955).

NITROBENZENE-^{15}N

$$(*NH_4)_2SO_4 \xrightarrow[\text{2. } H_2SO_4]{\text{1. } KMnO_4, 180°} H*NO_3 \xrightarrow{C_6H_6} Ph*NO_2$$

Procedure[1]

Ammonium chloride (0.537 g, 10 mmol) is dissolved in sulfuric acid (18 *M*, 0.5 ml) (Note 1). When the reaction is complete, water (10 ml) and sodium hydroxide (2 *M*) are added to adjust the pH to 3. Potassium permanganate (4.35 g, 27.5 mmol) is added, and the solution is heated in an autoclave at 180° for 20 hr, filtered from precipitated manganese dioxide, and evaporated to dryness. Sulfuric acid (18 M, 2 ml) is added very slowly, and to the solution at 50° is added benzene (1.5 ml, 17 mmol) over 0.5 hr. Refluxing at 70° is continued for 1 hr, the mixture is cooled, water (0.5 ml) is added, and the solution is made alkaline with sodium hydroxide (2 M) and extracted with ether (4 × 10 ml). The extract is dried over calcium chloride and evaporated to afford the product (0.284 g, 23% yield).

Notes

1   The ammonium chloride is converted to the sulfate to avoid complications from oxidation of chloride ion.

References

[1]J. H. Hog, "Synthesis of o-, m- and p-Deuterio-, [^{15}N]- and mono[^{18}O] Nitrobenzene," *J. Labelled Compds.*, 7, 179-181 (1971).

NITROBENZENE-^{18}O$_1$

$$H_2*O \xrightarrow{NO_2BF_4} HONO*O + BF_3 + HF \xrightarrow{C_6H_6} PhNO*O$$

Procedure[1] (Note 1)

Water-^{18}O (0.212 g, 11 mmol) is slowly distilled onto nitronium tetrafluoroborate (1.578 g, 11.8 mmol) on a vacuum line (Note 2). The gases are distilled in small portions onto benzene (1.5 ml, 17 mmol). The solution is stirred on the vacuum line for 24 hr, then heated to 60° for 30 min, made alkaline with sodium hydroxide (2 M), and extracted with ether (5 × 3 ml). The extract is dried over calcium chloride and evaporated to give the product (0.534 g, 40% yield) (Note 3).

Notes

1   The synthesis of ^{18}O-nitrobenzene cannot be performed efficiently by the usual nitration with conc. sulfuric acid and ^{18}O-nitric acid owing to equilibria that result in exchange.
2   Formation of the gases is quite violent.
3   The mass spectrum shows no detectable amount of nitrobenzene-^{18}O$_2$.

References

[1]J. H. Hog, "Synthesis of o-, m-, and p-Deuterio-[^{15}N]- and mono[^{18}O] Nitrobenzene," *J. Labelled Compds.*, 7, 179-181 (1971).

ANILINE-^{15}N

Method I

$$PhCO*NH_2 \xrightarrow{\text{NaOCl}} Ph*NH_2$$

Procedure[1]

A solution of sodium hydroxide (7.5 g, 0.18 mol) in water (20 ml) is saturated with chlorine at $0°$ after addition of ice (25 g). To the stirred hypochlorite solution benzamide-^{15}N (2.99 g, 24.7 mmol) is added, and stirring is continued for 1 hr at $100°$. The mixture is extracted with ether (3 × 50 ml), and the extract is dried over sodium sulfate and evaporated. The residue is distilled under vacuum to give the product (1.288 g, 55% yield).

Method II

$$Na*N_3 \xrightarrow[\text{PPA}]{\text{PhCMeBuCOOH}} Ph*NH_2$$

Procedure[2]

To a stirred solution of 2-methyl-2-phenylhexanoic acid (2.07 g, 10 mmol) in polyphosphoric acid (40 g) at $50°$ is added sodium azide-^{15}N$_3$ (0.650 g, 10 mmol), and stirring is continued for 8 hr. The mixture is treated with crushed ice, and the resulting solution is washed with methylene chloride (3 × 50 ml), made alkaline with solid sodium hydroxide while ice is added to maintain the temperature below $25°$, and extracted with methylene chloride. The extract is washed with water, dried over magnesium sulfate, and evaporated at reduced pressure to give an oil (0.594 g, 64% yield) (Note 1). The oil is dissolved in ether and treated with hydrogen chloride to precipitate aniline-^{15}N hydrochloride (0.694 g, 58% yield) (Notes 2 and 3).

Notes

1 GC (silicone gum nitrile column) shows only the product peak.
2 When benzoic acid is used instead of the trisubstituted acid the yield is only 47%.
3 The stated yields for the Schmidt reactions are based on sodium azide and do not consider the unrecovered two-thirds of the isotope.

Other Preparations

By the procedure of Method II have been prepared 2-phenylethylamine-^{15}N (71% yield) from 2-phenylpropionic acid, pentylamine-^{15}N (72% yield) from hexanoic acid, and heptadecylamine-^{15}N (98% yield) from stearic acid[2] (Note 3).

References

[1]J. Volford and D. Banfi, "Synthesis of ^{15}N-Labeled 2-Substituted 2-Thiazolines and Analogous Thiazines," *J. Labelled Compds.*, **11**, 419-426 (1975).

[2]J. L. Rabinowitz, R. M. Palmere, and R. T. Conley, "The Synthesis of ^{15}N Labeled Aniline and Other Amines," *J. Labelled Compds.*, **9**, 141-143 (1973).

3-METHOXYANILINE-^{15}N

Procedure[1]

3-Methoxybenzamide-^{15}N (0.251 g, 1.65 mmol) is added to a hypobromite solution prepared 5 min previously by dropwise addition of bromine (0.216 g, 1.35 mmol) to an ice-cold solution of sodium hydroxide (0.262 g, 6.5 mmol) in water (5.0 ml). The suspension is stirred vigorously for 20 min at 0°, and the resulting yellow solution is slowly heated to 95° and maintained at that temperature for 1.5 hr. The mixture is cooled and extracted with ether. The ether extract is dried over sodium sulfate and evaporated to give the product as a dark oil (0.129 g, 63% yield) (Note 1).

Notes

1   The crude product shows no detectable impurities by NMR.

Other Preparations

A series of 20 substituted anilines has been prepared by this procedure.[1]

References

[1]T. Axenrod, P. S. Pregosin, M. J. Wieder, E. D. Becker, R. B. Bradley, and G. W. A. Milne, "Nitrogen-15 Nuclear Magnetic Resonance Spectroscopy. Substituent Effects on ^{15}N-H Coupling Constants and Nitrogen Chemical Shifts in Aniline Derivatives," *J. Am. Chem. Soc.*, **93**, 6536-6541 (1971).

N-(METHYL-^{13}C)ANILINE-2,3,4,5,6-d_5

$$C_6D_5NHTs + *CH_3I \xrightarrow[Me_2CO]{NaOH} C_6D_5NTs*CH_3$$

$$
\begin{array}{l}
\text{1. } NH_3(1), Na, Et_2O \\
\text{2. } NH_4Cl, H_2O \\
\text{3. } H_3PO_4 \\
\text{4. } KOH
\end{array}
\xrightarrow{\hspace{2cm}} C_6D_5NH*CH_3
$$

Procedure[1]

To a solution of *N*-(phenyl-d_5)-4-toluenesulfonamide (0.80 g, 2.9 mmol), sodium hydroxide (1 *N*, 3 ml), and acetone (8 ml) is added methyl-^{13}C iodide (0.215 ml, 3.45 mmol). After the reaction is stirred for 72 hr methyl iodide (0.04 ml) is added, and stirring is continued another 29 hr at ambient temperature. The acetone is evaported at 60°, and the mixture is extracted with ether. Evaporation of the ether gives the *N*-(methyl-^{13}C)-*N*-(phenyl-d_5)-4-toluenesulfonamide as a pale-yellow solid (870 mg, 100% yield) (Note 1).

To the sulfonamide (750 mg, 2.6 mmol) in liquid ammonia (25 ml) and ether (25 ml) is added sodium (400 mg, 17 mmol) in small portions. The ammonia evaporates on standing overnight, and the excess sodium is decomposed with a saturated solution of ammonium chloride. The solution is acidified with phosphoric acid, washed with ether, made basic with potassium hydroxide, and extracted with ether to afford the product as a brown liquid (213 mg, 66% yield) (Notes 1 and 2).

Notes

1 NMR data are given.

2 The product is 96% pure by GC analysis.

Other Preparations

By the same procedure *N*-(methyl-^{13}C)-2,4,4-trimethyl-2-pentanamine is prepared in 49% yield.[1]

References

[1]G. Chapelet-Letourneux and A. Rassat, "Synthesis of Amines and Nitroxide Radicals Labeled with Carbon-13 and Deuterium," *Bull. Soc. Chim. Fr.,* **1971**, 3216-3221.

N-(ETHYL-2-^{13}C)-2,4,4-TRIMETHYL-2-PENTANAMINE

$$
*CH_3CONHCMe_2CH_2CMe_3 \xrightarrow[Et_2O]{LiAlH_4} *CH_3CH_2NHCMe_2CH_2CMe_3
$$

Procedure[1]

N-(1,1,3,3-Tetramethylbutyl)acetamide-2-^{13}C (980 mg, 5.7 mmol) in ether (30 ml) is refluxed with lithium aluminum hydride (1.5 g, 39 mmol) for 68 hr. The mixture is treated with water (1.5 ml), sodium hydroxide (15%, 1.5 ml), and water (4.5 ml), and filtered. The ether solution is dried over sodium sulfate and evaporated to give the product as a yellow liquid (690 mg, 77% yield).

Other Preparations

By the same procedure are prepared N-(benzyl-1-^{13}C)-2,4,4-trimethyl-2-pentanamine, N-$tert$-butyl(benzyl-1-^{13}C)amine, and N-(phenyl-d_5)(benzyl-1-^{13}C)amine.

References

[1]G. Chapelet-Letourneux and A. Rassat, "Synthesis of Amines and Nitroxide Radicals Labeled with Carbon-13 and Deuterium," *Bull. Soc. Chim. Fr.,* **1971,** 3216-3221.

4-NITROANILINE-1-^{13}C AND -1,4-^{13}C$_1$

Procedure[1]

4-Nitroaniline-1-^{13}C 4-Nitrophenol-1-^{13}C (1.58 g, 11.4 mmol), 4-chloro-2-phenyl-quinazoline (3.37 g, 14.0 mmol), and potassium carbonate (3.30 g, 23.9 mmol) in acetone (250 ml) are refluxed under nitrogen (Note 1) for 27 hr and then poured onto ice (225 g). The pale-yellow adduct is filtered and recrystallized from benzene (200 ml); mp 220-220.5° (3.74 g, 96% yield).

The neat adduct (1.81 g) is heated in a Wood's metal bath at 285-295° for a few minutes, cooled, examined visually and by IR, and the process is repeated. After a total 25-min thermolysis time (Note 2), the flask and contents are reduced to fine particles and treated with methanol (95%, 70 ml), water (5.5 ml), and potassium hydroxide (5.5 g) (Note 1). The mixture is refluxed under nitrogen for 1.5 hr, acidified with hydrochloric acid (12 M), transferred with the aid of deaerated methanol (25 ml) to a clean flask through a narrow stem funnel (Note 3), stirred at 51-52° for 30 min, cooled in an ice bath, and made basic with potassium hydroxide (saturated aqueous solution). The mixture containing a yellow precipitate is concentrated at reduced pressure to a mushy dryness, treated with water (20 ml), and extracted with ether (5 × 100 ml) (Note 4). The extract is dried and evaporated, and the yellow residue is purified by chromatography on an alumina column with hexane-ether to give the product (obtained with 80% ether), mp 145-146.5° (247 mg).

Repetition of the procedure with the other portion of adduct (1.90 g) gives additional product (234 mg) (Note 5). The recovered starting material (932 mg) from the two runs is recycled through the procedure (total yield 955 mg, 69%, Note 6).

4-Nitroaniline-1,4-^{13}C$_1$ 4-Nitroaniline-1-^{13}C (932 mg, 6.75 mmol) is oxidized with trifluoroperacetic acid[2] to 1,4-dinitrobenzene-1-^{13}C, mp 170-171.5° (955 mg, 84% yield). The latter compound (563 mg, 3.65 mmol) is subjected to a slight modification of the published procedure.[3] The reactants are stirred at 85° for 1 hr and then at room temperature for 1.5 hr. The ether extracts (6 × 200 ml) of the reaction mixture are concentrated to approx. 75 ml and washed several times with hydrochloric acid (10%). The aqueous layer is made basic with sodium hydroxide (50%) and extracted with ether (8 × 125 ml). The extract is dried and evaporated to give the crude product, which is purified by column chromatography (florisil, 25% ether-hexane) and TLC (silica gel, methylene chloride); mp 144.5-145.5° (368 mg, 80% yield).

Notes

1   As a precaution against the possibility of free-radical decomposition products, solvents are deaerated, and reactions are run under a nitrogen atmosphere.

2   The times, from liquefaction until removal from the heat, were 4, 4, 7, 5, and 5 min.

3   No paper is used, thereby allowing transfer of the precipitate but removal of glass fragments.

4   The aqueous solution is acidified and extracted with ether (6 × 200 ml) for recovery of starting material (451 mg).

5   Starting material (481 mg) is recovered.

6   The amount of starting material consumed is 1.40 g.

References

[1] G. L. Swartz and W. M. Gulick, Jr., "Synthesis of Carbon-13 Enriched Nitrobenzene," *J. Labelled Compds.*, **11**, 525-534 (1975).

[2] A. S. Pagano and W. D. Emmons, "2,6-Dichloronitrobenzene," *Org. Syn.*, **49**, 47-50 (1969).

[3] W. W. Hartman and H. L. Silloway, "2-Amino-4-nitrophenol," *Org. Syn.*, Col. Vol. 3, 82-84 (1955).

N,N-DIMETHYLANILINE OXIDE-^{18}O

$$*O_2 \xrightarrow[\text{H}_2\text{O}]{\text{Ba(OH)}_2} Ba*O_2 \xrightarrow[\text{HOAc, H}_2\text{SO}_4]{\text{PhNMe}_2} PhN(*O)Me_2$$

Procedure[1]

Barium peroxide-$^{18}O_2$ Oxygen-$^{18}O_2$ is transferred to a 500-ml flask containing 60 ml of a suspension of barium hydroxide (Note 1), which is stirred for 48 hr. The mixture is centrifuged, and the precipitate is washed with water (2 × 60 ml), acetone, and ether; (204 mg, approx. 40% yield, Note 2).

N,N-Dimethylaniline oxide-^{18}O hydrochloride Barium peroxide-$^{18}O_2$ octahydrate (204 mg, 0.65 mmol) is introduced with cooling into acetic acid (1.25 ml) (Note 3). After complete solution, sulfuric acid (74 mg, 0.75 mmol) in acetic acid (0.25 ml) is added. The precipitated barium sulfate is centrifuged and washed with acetic acid (2 × 0.5 ml). To the combined clear supernatants at 4° is added N,N-dimethylaniline (121 mg, 1 mmol), and the mixture is kept at 4° for 4 days. Acetic acid and excess amine are distilled under vacuum at the lowest possible temperature, and the pale-violet residue is taken up in water (10 ml). The solution is extracted twice with ether, the ether extract is washed with water, and the aqueous solution is treated dropwise with sufficient barium chloride solution (0.1 M) or sulfuric acid (0.1 M) until barium sulfate no longer precipitates. Hydrochloric acid (1 N, 2 ml) is added, and the supernatant is evaporated. The residue is dried over potassium hydroxide and calcium chloride and dissolved in acetone (2 × 10 ml) at 30°. The solution is evaporated, and the residue is recrystallized from hot acetone with slow cooling to −15° to afford the product as long needles, mp 121-122° (43.5 g, 38% yield).

Notes

1 Sufficient precipitated barium hydroxide is present to ensure saturation of the solution throughout the reaction.
2 The product is the octahydrate of 80-85% purity.
3 Caution.

References

[1]H.-L. Schmidt, N. Weber, and M. Holmann, "The Labeling of Amine Oxides with Isotopic Oxygen," *J. Labelled Compds.*, 7, 171-174 (1971).

N-(NITROSO-^{15}N)DIPHENYLAMINE-^{15}N

$$Ph*NH_2 \xrightarrow[\text{HCl, CCl}_4]{\text{Ac}_2\text{O}} Ph*NHAc \xrightarrow[\text{K}_2\text{CO}_3, \text{PhNO}_2]{\text{PhI, Cu}_2\text{I}_2}$$

$$Ph_2*NAc \xrightarrow[\text{EtOH}]{\text{HCl}} Ph_2*NH \xrightarrow[\text{H}_2\text{O, C}_6\text{H}_6]{\text{Na*NO}_2, \text{HCl}} Ph_2*N*NO$$

Procedure[1]

Acetanilide-^{15}N A solution of aniline-^{15}N (1.00 g, 10.6 mmol) in carbon tetrachloride (15 ml) is refluxed with hydrochloric acid (conc., 1 ml) and acetic anhydride (5 ml) for 30 min. Methanol (15 ml) is added, and the methyl acetate formed is distilled. Carbon tetrachloride is added, distilled to azeotrope water, and evaporated to give the product as a flaky residue, mp 111-112° (1.43 g, 99% yield).

N-Acetyldiphenylamine-^{15}N Acetanilide-^{15}N (1.02 g, 7.5 mmol), iodobenzene (2.25 g, 11 mmol), potassium carbonate (0.6 g), and cuprous iodide (15 mg) are refluxed in nitrobenzene (4 ml) for approx. 20 hr. Nitrobenzene and excess iodobenzene are removed by steam distillation. The oily residue is extracted with ether, and the extract is dried and evaporated to give the crystalline product (1.43 g, 90% yield).

Diphenylamine-^{15}N N-Acetyldiphenylamine (1.43 g, 6.74 mmol) is refluxed with alcohol (5 ml) and hydrochloric acid (conc., 4 ml) for 3 hr. Steam distillation (400 ml distillate) affords the crystalline product, mp 46-48.5° (1.05 g, 92% yield) (Note 1).

N-(Nitroso-^{15}N)diphenylamine-^{15}N Diphenylamine-^{15}N (0.85 g, 5.0 mmol) in benzene (10 ml) (Note 2) and sodium nitrite (0.5 g, 7 mmol) in water (5 ml) are stirred vigorously at approx. 5°, and hydrochloric acid (conc., 1.5 ml) is added dropwise over 30 min. The mixture is stirred another hour and extracted with benzene (60-70 ml). The extract is washed with water, dried, and evaporated under nitrogen to give the product (0.98 g, 99% yield), which is recrystallized from pentane and alcohol; mp 62.5-63.5° (Note 3).

Notes

1 The overall yield from aniline is 82%.
2 Because diphenylamine hydrochloride is not very soluble in water, benzene is used and gives better yields than alcohol.
3 IR and NMR data are given.

References

[1]S. Bulusu and J. R. Autera, "Synthesis of ^{15}N-Labeled Diphenylamine and Diphenylnitroso-amine," *J. Labelled Compds.*, **10**, 511-514 (1974).

TOLUENE-1-^{13}C OR -α-^{13}C

Procedure

Toluene-1-[13]C[1]   A 10-ml pear-shape flask containing 1-methylcyclohexene-1-^{13}C is attached to a 30-cm column that is loosely packed with palladium-on-charcoal (30%, approx. 20 g) on top of a glass-wool plug and heated to 400° with heating tape. The flask is heated in an oil bath to 140-170°, and the distillate (after passing through the catalyst column) is condensed in a Dry Ice-cooled receiver. This procedure is repeated three times (Note 1). Following the fourth pass, the apparatus is evacuated (0.5 mm Hg) to remove residual product from the column; (1.30 g, 74% yield).

Toluene-α-[13]C[1]   By the same procedure 1-(methyl-^{13}C)cyclohexene is converted to this isotope isomer (1.72 g, 89% yield) (Note 2).

α-Bromotoluene-1-[13]C[1]   A solution of toluene-1-[13]C (1.25 g, 13 mmol) in carbon tetrachloride (25 ml) is irradiated with a 200-watt incandescent lamp while bromine (2.15 g, 13 mmol) in carbon tetrachloride (12.5 ml) is added at such a rate that the color of the reaction remains pink until addition is completed. Irradiation is continued an additional minute, and water (20 ml) is quickly added. The organic layer is separated, washed with water (3 × 25 ml), dried over calcium chloride, and evaporated. The residue is distilled under vacuum to give the product, bp 52-56° at 0.5 mm Hg (1.49 g, 64% yield).

Notes

1   The flask is recharged with the distillate after each run. The reaction is monitored by ^{13}C-NMR, which shows carbon-13 at C-1 only.
2   GC and NMR show the product to be greater than 99% pure.

Other Preparations

Toluene-1-^{13}C has been prepared in 60% yield from 1-methylcyclohexene-1-^{13}C using Shell 105 Dehydrogenation Catalyst at 510°.[2] Dehydrogenation of the cyclohexene using 20% palladium-on-carbon catalyst at 400° has given toluene-1-^{13}C in 70% yield.[3]

References

[1] G. A. Braden and U. Hollstein, "Synthesis of 1-Phenyl-2-phenyl-1-^{13}C-ethene-1-[13]C (*trans*-Stilbene) and Derivatives," *J. Labelled Compds.*, **12**, 507-516 (1976).
[2] B. Stuetzel, W. Ritter, and K-F. Elgert, "^{13}C-Enriched Monomers for NMR-Spectroscopy of Polymers," *Angew. Makromol. Chem.*, **50**, 21-41 (1976).
[3] R. J. Dickinson and D. Williams, "Investigation of Possible Isomerization Reactions in Benzene, Iodobenzene, and Phenol upon Electron Impact," *J. Chem. Soc. B*, **1971**, 249-251.

BENZYLAMINE-¹⁵N HYDROCHLORIDE

Method I

$$*NH_3 + PhCHO \xrightarrow[\text{MeOH}]{H_2O} PhCH=*NH$$

$$\xrightarrow[\substack{\text{1. NaCNBH}_3 \\ \text{HOAc, MeOH} \\ \text{2. NaOH} \\ \text{3. HCl}}]{} PhCH_2*NH_2 \cdot HCl$$

Procedure[1]

Benzylimine-¹⁵N   Equal volumes of ammonia-¹⁵N solution (approx. 10 M, in water) and benzaldehyde solution (approx. 10 M, in methanol) are mixed in a low, wide beaker and placed in a small desiccator well-packed with potassium hydroxide pellets (Note 1). When all the solvent has been absorbed, the crystals of the relatively unstable product (decomp. 93-96°) are collected and stored under cool, dry conditions (Note 2).

Benzylamine-¹⁵N hydrochloride   To benzylimine-¹⁵N (1.06 g, 10 mmol) is added methanol (10 ml) containing acetic acid (0.6 g, 10 mmol) and sodium cyanoborohydride (0.88 g, 14 mmol). After 1 hr the solvent is removed, sodium hydroxide (5 N) is added, and the solution is repeatedly extracted with ether. The extract is extracted with hydrochloric acid (1 N), and the aqueous solution is filtered and concentrated. The residue is recrystallized from methanol-hexane and from methanol to give the product, mp 254-256° (69% yield from ammonia-¹⁵N).

Method II

Procedure[2]

Benzylamine-¹⁵N   Benzamide-¹⁵N (1.50 g, 12.3 mmol) is placed in a filter thimble that is inserted into a 150-ml cylindrical, pressure-equalizing funnel to which a reflux condenser and drying tube are attached. Lithium aluminum hydride (1 g, 26 mmol) and ether (250 ml) are refluxed for 4 hr, allowing the ether to carry the amide slowly into the reaction medium. The mixture is stirred for another 12 hr at room temperature and treated with water (1 ml), sodium hydroxide (15%, 1 ml), and water (3 ml). The readily filtered cake of aluminum hydroxide thus formed is washed with ether, and the ether solution is evaporated at low temperature to afford the product (Note 3).

Notes

1 High yields are obtained only if the air space is kept small. Well-formed white crystals appear within hours.
2 The best yield is 97%. Treatment with hydrogen chloride in ether gives benzyliminium-^{15}N chloride, mp 179-181°.
3 Yields of approx. 80% are obtained in similar runs with unlabeled material.

References

[1]C. Gazzola and G. L. Kenyon, "A Two-Step Synthesis of Benzylamine-^{15}N from Ammonia-^{15}N," J. Labelled Compds., 15, 181-184 (1978).

[2]U. Hornemann, "Synthesis of 2-amino-2-deoxy-D-glucose-^{15}N and of 2-amino-2-deoxy-L-glucose-2-^{14}C," Carbohydr. Res., 28, 171-174 (1973).

N-BENZYL-N-(2-CHLOROETHYL)-1-PHENOXY-2-PROPANAMINE-^{15}N HYDROCHLORIDE (^{15}N-PHENOXYBENZAMINE)

$$PhOCH_2CH(CH_3)*N(CH_2Ph)CH_2CH_2OH \quad \xrightarrow[\text{2. SOCl}_2]{\text{1. HCl, CHCl}_3}$$

$$PhOCH_2CH(CH_3)*N(CH_2Ph)CH_2CH_2Cl \cdot HCl$$

Procedure[1]

2-[Benzyl(1-methyl-2-phenoxyethyl)amino]ethanol-^{15}N (2.9 g, 10.2 mmol) in chloroform (5 ml) is treated with hydrogen chloride (Note 1) and cooled to 0°, and a solution of thionyl chloride (1.21 g, 10.2 mmol) in chloroform (5 ml) is added dropwise. After the reaction warms to room temperature, it is heated for 2 hr at 50-60°, and the solvent is removed. The resulting tan oil is triturated with ether to give a white solid, which is recrystallized twice from chloroform-ether, mp 136.5-138° (1.1 g, 32% yield). Additional product of equal purity (Note 2) is isolated from the mother liquors (total yield 1.65 g, 48%).

Notes

1 The apparent pH to wet test paper is 1-2.
2 TLC, UV, and other analytical data are given.

References

[1]W. L. Mendelson, L. E. Weaner, L. A. Petka, and D. W. Blackburn, "The Preparation of ^{15}N-Phenoxybenzamine," J. Labelled Compds., 11, 349-353 (1975).

N-PHENYL-2-NAPHTHYLAMINE-8-^{13}C

Procedure[1]

A mixture of 2-naphthol-8-^{13}C (613 mg, 4.25 mmol), aniline (790 mg, 8.5 mmol), and 4-toluenesulfonic acid (10 mg) is heated under nitrogen for 1 hr at 190° and then for 6 hr at a temperature slowly rising to 240° (Note 1). A solution of the cooled melt in acetone-hexane (1:1) is applied to a silica gel column (100 mesh, 10-mm diam × 8-cm long) and eluted with acetone-hexane (1:9) (Note 2). Dropwise addition of water to a warm acetic acid (3 ml) solution of the evaporated eluate (925 mg) produces fawn needles, which are dissolved in ethyl acetate, washed with sodium hydroxide (1 N) and water, dried, and recrystallized from ethyl acetate (606 mg, 65% yield) (Note 3).

Notes

1 The reaction vessel facilitates escape of the water produced.
2 Elution is continued until GC analysis indicates product is no longer present in the eluate.
3 GC and TLC analyses show the product to be pure and free of the 2-naphthyl-amine-8-^{13}C that is present in the crude product.

References

[1]G. H. Walker and D. E. Hathaway, "Synthesis of N-Phenyl-2-[1,4,5,8-^{14}C] naphthylamine, N-Phenyl-2-[8-^{13}C] naphthylamine, and N-[U-^{14}C] Phenyl-2-naphthylamine," *J. Labelled Compds.*, **12**, 190-206 (1976).

DIPHENYLMETHANIMINE-^{15}N

Procedure[1]

Benzonitrile-^{15}N (1.87 g, 18.0 mmol) in ether (10 ml) is added dropwise at room temperature to phenyl Grignard reagent prepared from bromobenzene (3.16 g, 20.1 mmol), and the mixture is stirred for 7.5 hr. The solution is cooled, methanol (3.61 g) is added carefully at room temperature, and the resulting gum is stirred for 30 min until completely crystalline. The slurry is filtered, the solvents are distilled, and the residue on distillation gives the product, bp 120° at 1 mm Hg (2.16 g, 67% yield).

Other Preparations

By the same procedure are prepared 5-nonanimine-^{15}N (41% yield) and 2-methyl-1-phenylbutanimine-^{15}N (74% yield).[1]

References

[1] J. B. Lambert, W. L. Oliver, and J. D. Roberts, "Nitrogen-15 Magnetic Resonance Spectroscopy. IV. The Degenerate Bimolecular Exchange of Protons in Ketimines," *J. Am. Chem. Soc.*, 87, 5085-5090 (1965).

BENZIDINE-^{15}N$_2$

$$\text{H}_2\text{*NCO} - \langle \bigcirc \rangle - \langle \bigcirc \rangle - \text{CO*NH}_2 \quad \xrightarrow[\substack{3.\ \text{Na}_2\text{SO}_4, \text{H}_2\text{O} \\ 4.\ \text{NaOH}}]{\substack{1.\ \text{NaOCl, NaOH,}\ \text{dioxane} \\ 2.\ \text{HCl}}}$$

$$\text{H}_2\text{*N} - \langle \bigcirc \rangle - \langle \bigcirc \rangle - \text{*NH}_2$$

Procedure[1]

A mixture of 4,4'-biphenyldicarboxamide-^{15}N$_2$ (4.84 g, 20 mmol), dioxane (200 ml), and an aqueous sodium hypochlorite solution (240 ml, approx. 60 mmol) (Note 1) is heated to 80° for 3 hr. The reddish-brown mixture is cooled, acidified with hydrochloric acid (10%, 200 ml), treated with Norit, and filtered through Celite. Treatment of the filtrate with sodium sulfate (1 M, 100 ml) precipitates crude benzidine-^{15}N$_2$ sulfate, which is filtered, washed with water, ethanol, and ether, and dried to give a tan powder (4.22 g). The solid is stirred with sodium hydroxide (10%, 100 ml) and ether (300 ml) for 30 min. The ether layer is separated, the aqueous layer is extracted with ether (3 × 200 ml), and the combined extracts are dried over magnesium sulfate and evaporated. The residue is recrystallized from aqueous ethanol to give the product as a light-tan powder, mp 124-125° (1.66 g, 45% yield).

Notes

1 The hypochlorite solution is prepared by dissolving the chlorine generated from potassium permanganate (4.86 g, 30 mmol) and hydrochloric acid (conc., 63 ml, 750 mmol) in sodium hydroxide (10%, 300 ml).

References

[1]T. W. Whaley, "Syntheses with Stable Isotopes: Benzidine-$^{15}N_2$," *J. Labelled Compds.*, **14**, 243-248 (1978).

DODECANE-1,12-$^{13}C_2$

$$HO*CH_2(CH_2)_{10}*CH_2OH \xrightarrow[\text{C}_5\text{H}_5\text{N}]{\text{TsCl}} TsO*CH_2(CH_2)_{10}*CH_2OTs$$

$$\xrightarrow[\text{Et}_2\text{O}]{\text{LiAlH}_4} *CH_3(CH_2)_{10}*CH_3$$

Procedure[1]

1,12-Dodecanediyl-1,12-$^{13}C_2$ bis(4-toluenesulfonate) To 1,12-dodecanediol-1,12-$^{13}C_2$ (2.5 g, 12 mmol) in pyridine (40 ml) at 0° is added *p*-toluenesulfonyl chloride (8.0 g, 42 mmol), and the solution is allowed to stand for 24 hr at 0°. Ice (approx. 100 g) is added, the mixture is stirred for 15 min, and the precipitate is filtered, washed with water, and recrystallized from ethanol as plates, mp 90-91° (3.15 g, 80% yield).

Dodecane-1,12-$^{13}C_2$ The ester above (3.0 g, 5.8 mmol) is added at room temperature to a stirred suspension of lithium aluminum hydride (1.0 g, 26 mmol) in ether (50 ml), and the mixture is stirred at reflux for 2 days, cooled, and treated with dilute hydrochloric acid. The ether layer is removed, dried over sodium sulfate, and evaporated at reduced pressure to give an oil (0.7 g), which is purified by distillation (bp 212-216°) and column chromatography (silica gel, petroleum ether) to give the product as a colorless oil (0.3 g, 30% yield).

References

[1]G. J. Shaw and G. W. A. Milne, "Synthesis and Properties of Dodecane-1,12-$^{13}C_2$ and Dodecane-1,1,1,12,12,12-$^2H_6$," *J. Labelled Compds.*, **12**, 557-563 (1976).

$^{13}C_2$-1,2-DIPHENYLETHANE

Procedure[1]

1-Phenyl-2-(phenyl-1-^{13}C)ethene-1-^{13}C (^{13}C$_2$-*trans*-stilbene) To a stirred solution of sodium methoxide (0.285 g, 5.2 mmol) and benzyl-1-^{13}C diethylphosphonate (1.005 g, 4.6 mmol) in dimethylformamide (5 ml) is added dropwise benzaldehyde-^{13}C (0.468 g, 4.4 mmol) in dimethylformamide (5 ml). The temperature is kept between 30 and 40° by stirring and cooling over ice when necessary. The solution is allowed to cool to room temperature and poured onto ice. The white crystals are filtered, washed twice with water, dried, and recrystallized from ethanol; mp 127-128° (0.837 g, 60% yield).

1-Phenyl-2-(phenyl-1-^{13}C)ethane-1-^{13}C A solution of the stilbene above (48.1 mg, 0.266 mmol) in ether (25 ml) and platinum oxide (0.02 g, 0.09 mmol) is hydrogenated at approx. 45 psi for 0.5 hr on a Parr shaker. The solution is filtered and evaporated, and the crystals are recrystallized from methanol; mp 55-56.5° (41.4 mg, 85% yield).

References

[1]G. A. Braden and U. Hollstein, "Synthesis of 1-Phenyl-2-phenyl-1-^{13}C-ethene-1-^{13}C (*trans*-Stilbene) and Derivatives," *J. Labelled Compds.*, **12**, 507-516 (1976).

^{13}C$_2$-TOLAN

Procedure[1]

***meso*-1,2-Dibromo-1-phenyl-2-(phenyl-1-^{13}C)ethane-1-^{13}C** To a stirred solution of the ^{13}C$_2$-stilbene (45.6 mg, 0.25 mmol) in ether is added bromine (48 mg, 0.30 mmol), and stirring is continued for 1 hr. Th crystals are filtered and washed with ether until white, mp 218-220° (59.7 mg, 69% yield).

1-Phenyl-2-(phenyl-1-^{13}C)ethyne-1-^{13}C To a solution of potassium hydroxide (76 mg, 1.4 mmol) in ethanol (2 ml) at just below reflux temperature is added the dibromo compound above (59.7 mg, 0.175 mmol), and the mixture is refluxed for 24 hr. The solvent is removed, and the product is recrystallized from ethanol; mp 63.3-64.5° (20.8 mg, 65% yield).

References

[1]G. A. Braden and U. Hollstein, "Synthesis of 1-Phenyl-2-phenyl-1-^{13}C-ethene-1-^{13}C (*trans*-Stilbene) and Derivatives," *J. Labelled Compds.*, **12**, 507-516 (1976).

o-TERPHENYL-2-^{13}C

Procedure[1]

2-(1-Cyclohexenyl-2,6-^{13}C$_1$)biphenyl To a Grignard solution from 2-iodobiphenyl (9.4 g, 33.5 mmol) and magnesium (0.96 g, 40 mmol) in ether (30 ml) is added dropwise with ice-bath cooling crude cyclohexanone-2-^{13}C (1.64 g, 16.7 mmol). The mixture is stirred at room temperature 1 hr and at reflux 1 hr and then treated with saturated ammonium chloride solution. The ether solution is separated, the aqueous phase is extracted with ether (200 ml), and the combined ether extracts are washed with sodium bicarbonate solution, dried over sodium sulfate, and evaporated. The residue (7.0 g) is sublimed for 10 hr at 140° (Note 1) to leave a residue of crude product (2.29 g, 59% yield).

o-Terphenyl-2-^{13}C The above-mentioned sublimation residue (2.29 g, 9.8 mmol) (Note 2) is heated with palladium-on-carbon (10%, 4.5 g) at 270° (Note 3) for 10 hr with magnetic stirring. The mixture is extracted with benzene (100 ml total) to give a solution of the crude product (1.66 g, 73% yield, Note 4), which is further purified by preparative thin-layer chromatography (silica gel, hexane) and precipitation from methanol-water; mp 54-56° (87% recovery).

Notes

1 The sublimation removes the biphenyl and also dehydrates the alcohol.
2 The use of starting material containing carbinol, or pure carbinol, gives essentially the same yield of product.
3 At higher temperatures increased amounts of by-product triphenylene are formed.
4 About 20% by-products are also present.

References

[1]F. Geiss and G. Blech, "Synthesis of *o*-Terphenyl-2-^{13}C," *J. Labelled Compds.*, 4, 119-126 (1968).

$^{13}C_1$-BENZO[*a*]PYRENE

Method I

Procedure[1]

Ethyl 2-(1-hydroxy-1,2,3,4-tetrahydrobenz[*a*]anthracen-1-yl)acetate-1-^{13}C or ethyl 2-(1-hydroxy-1,2,3,4-tetrahydrobenz[*a*]anthracen-1-yl)acetate-2-^{13}C   Butyllithium in hexane (1.6 *M*, 9.4 ml, 15 mmol) is added to *N*-isopropylcyclohexylamine (2.33 g, 16.5 mmol) in tetrahydrofuran under nitrogen at −78°. To this mixture is added ethyl acetate-1-^{13}C or ethyl acetate-2-^{13}C (1.33 g, 15.0 mmol) in tetrahydrofuran (15 ml) at a rate that maintains the temperature below −75°. Stirring is continued another 15 min, and 3,4-dihydrobenz[*a*]anthracen-1(2*H*)-one (3.69 g, 15.0 mmol) in tetrahydrofuran (45 ml) is added at a rate that maintains the temperature below −75°. Stirring is continued at −78° for 1 hr, and the orange complex is hydrolyzed by the dropwise addition of hydrochloric acid (conc., 2 ml) in tetrahydrofuran (10 ml) at a rate that maintains the temperature below −70° (Note 1). The mixture is allowed to warm to room temperature, water and ether (50 ml each) are added, the ether layer is washed with hydrochloric acid (5%, 2 × 20 ml), and the aqueous layer is extracted with ether. The combined ether extracts are dried over magnesium sulfate and evaporated to afford an orange oil that solidifies on trituration with ethanol (Note 2). The solid is crystallized from ethanol (95%) to give pale-yellow crystals, mp 114.5-115.5° (4.02 g, 80% yield).

Ethyl 1-Benz[a]anthraceneacetate-α-^{13}C or ethyl 1-benz[a]anthraceneacetate-*carboxy*-^{13}C In a dehydrogenation tube fitted with a ground-glass cold finger condenser and gas inlet and outlet tubes are placed the above hydroxy ester (2.00 g, 6.0 mmol), palladium-on-charcoal (10%, 0.20 g), 1,1-diphenylethene (1.20 g, 6.6 mmol), and 1-methylnaphthalene (10 ml). The mixture is placed in a preheated Wood's metal bath and heated at 250-260° for 2 hr while steam is passed through the condenser and a slow flow of nitrogen is maintained. The reaction mixture is diluted with benzene and filtered, and the benzene is evaporated. 1-Methylnaphthalene, 1,1-diphenylethane, and 1,1-diphenylethene are removed on a Kugel-Rohr at 50-60° and 0.25 mm Hg. The orange oil is crystallized from ethanol (95%) to afford the product as beige needles, mp 107-109.5° (1.54-1.67 g, 82-89% yield) (Note 3).

Benzo[a]pyrene-11-^{13}C or benzo[a]pyrene-12-^{13}C A solution of the ester above (1.42 g, 4.5 mmol) in toluene (45 ml) at −75° under nitrogen is treated with diisobutylaluminum hydride in hexane (1 *M*, 4.5 ml, 4.5 mmol). The mixture is stirred for 1 hr at −75°, and the pale-yellow complex is hydrolyzed with hydrochloric acid (conc., 1 ml) in tetrahydrofuran (9 ml). The mixture is warmed to room temperature and washed with ammonium chloride (5%), and the toluene layer is dried over magnesium sulfate and evaporated to give the aldehyde as a pale-yellow oil. The aldehyde is dissolved in methanesulfonic acid (65 ml) and stirred under nitrogen for 40 min on a warm water bath. The deep-red reaction mixture is poured into water and ice (100 ml), and the precipitated crude yellow product is chromatographed (neutral alumina, benzene). The benzene eluate is concentrated, and methanol is added to give shiny platelets, mp 177.5-178° (0.70 g). From the mother liquor is obtained additional product, mp 176-177° (total yield 0.93 g, 82%).

Method II

1. Li•CH$_2$*COOEt,
   −75°, THF

2. HCl, −70°

Pd/C, 250°
Ph$_2$C=CH$_2$
C$_{10}$H$_7$Me

•CH$_2$*COOEt

1. (i-Bu)$_2$AlH,
   PhCH$_3$, −75°
2. HCl, H$_2$O, THF
3. MeSO$_3$H
4. H$_2$O

Procedure[1]

Ethyl 4-hydroxy-1,2,3,4-tetrahydro-4-chryseneacetate-α-^{13}C or ethyl 4-hydroxy-1,2,3,4-tetrahydro-4-chryseneacetate-*carboxy*-[13]C By the procedure of Method I, for the hydroxy ester, 1,2-dihydrochrysen-4(3*H*)-one (4.92 g, 20.0 mmol) and the lithium enolate from ethyl acetate-1-^{13}C or ethyl acetate-2-[13]C (1.78 g, 20.0 mmol) give the product as colorless crystals, mp 77-81° (5.42 g, 82% yield) (Note 4).

Ethyl 4-chryseneacetate-α-^{13}C or ethyl 4-chryseneacetate-*carboxy*-[13]C Dehydration and dehydrogenation of the hydroxy ester above (2.5 g, 7.5 mmol) as described for the analogous compound in Method I gives colorless prisms, mp 63.5-65° (1.75 g, 74% yield).

Benzo[a]pyrene-4-^{13}C or benzo[a]pyrene-5-[13]C Reduction of the above-metioned ester (0.94 g, 3.0 mmol) by the method described for the analogous compound in Method I followed by cyclization gives a pale-green solid, which is chromatographed (neutral alumina) and recrystallized from benzene-methanol to afford the product, mp 176.5-178° (0.64 g, 85% yield).

Notes

1 The hydroxy ester undergoes rapid reversion to the ketone if the mixture is allowed to warm before acidification.
2 Trituration of the crude product with cold hexane has been found in subsequent runs to afford good product, mp 114-116° (82% yield).
3 Recrystallization from 95% ethanol gives mp 109-109.5°.
4 Further recrystallization failed to improve the melting point or to remove a trace of the starting ketone as shown by TLC.

References

[1]R. S. Bodine, M. Hylarides, G. H. Daub, and D. L. Vander Jagt, "[13]C-Labeled Benzo[a]pyrene and Derivatives. 1. Efficient Pathways to Labeling the 4,5,11, and 12 Positions," *J. Org. Chem.*, 43, 4025-4028 (1978).

6-(METHYL-^{13}C)BENZO[a]PYRENE

$$H_2NNH_2, KOH$$
$$(HOCH_2CH_2)_2O$$
$$100\text{-}150°$$

*CHO   *CH_3

Procedure[1]

6-Benzo[a]pyrenecarboxaldehyde-formyl-[13]C (0.7 g, 2.5 mmol), hydrazine hydrate (0.66 ml, 13 mmol), diethyleneglycol (40 ml), and potassium hydroxide (0.44 g, 7.9 mmol) are heated with stirring for 30 min at 100° and then for 2 hr at 140-150°. The mixture is cooled, water is added, and the precipitate is filtered, dried, and chromatographed (alumina, benzene). The eluant is concentrated and treated with ethanol to crystallize the product, mp 214-215° (0.42 g, 60% yield). Sublimation and recrystallization from benzene-ethanol gives material with mp 215-216°.

References

[1]R. E. Royer, G. H. Daub, and D. L. Vander Jagt, "Synthesis of Carbon-13 Labelled 6-Substituted Benzo[a]pyrenes," J. Labelled Compds., 12, 377-380 (1976).

5-[3-(METHYLAMINO)PROPYLIDENE-3,3-d_2]-10,11-DIHYDRO-5H-DIBENZO[a,d]CYCLOHEPTENE-[15]N (d_2-NORTRIPTYLINE-^{15}N)

Procedure[1]

N-Methyl-3-(10,11-dihydro-5H-dibenzo[a,d]cycloheptylidene)propanamide-[15]N
The acid chloride is prepared from the acid (660 mg, 2.5 mmol) and oxalyl chloride (0.5 ml, 2.9 mmol) in benzene (24 ml). Reaction with methylamine-^{15}N hydrochloride (240 mg, 1.5 mmol), sodium hydroxide (5 N, 3 ml), and water (1 ml) gives the crude cream colored product, mp 125-130° (435 mg). Preparative thin layer chromatography (silica gel, ethyl acetate-methylcyclohexane, 2:1) affords the product, mp 132-134° (330 mg, 48% yield).

d_2-Nortriptyline-[15]N   To sodium borodeuteride (84 mg, 2.0 mmol) and the amide (140 mg, 0.5 mmol) in tetrahydrofuran (3 ml) in a nitrogen-swept flask cooled in ice water is added, all at once, dimethyl sulfate (0.19 ml, 2 mmol) in tetrahydrofuran (2 ml) (Note 1). The mixture is stirred until it reaches room temperature, warmed gently for 2 hr, refluxed for 8 hr, cooled, and hydrolyzed with deuterium chloride (38%) in water-d_2. It is heated for 2 hr (most of the tetrahydrofuran boils off), made strongly basic, and extracted with ether to afford the product, mp 211-213° (20% yield) after purification by thin layer chromatography.

Notes

1 The simple generation of the selective reducing agent diborane from sodium borohydride and methyl sulfate is also applicable for the preparation of non-deuterated compounds and is particularly suited where use of lithium aluminum hydride is unsuccessful (as in this case) or would lead to removal of an aromatic halogen substituent.

Other Preparations

$^{13}C_2$-Nortriptyline, [5-(3-methylaminopropylidine-3-^{13}C)-10,11-dihydro-5H-dibenzo[a,d] cycloheptene-5,10-$^{13}C_1$], $^{13}C_2$-desmethylnortriptyline, and $^{13}C_2$-amitriptyline (the analogous amino and dimethylamino compounds, respectively) have been prepared using previously developed procedures (involving some 13 steps) starting from phthalic-$^{13}C_1$ acid. The single label is distributed among positions 5, 10, and 11 in the tricyclic ring; the second carbon-13 is introduced in the side chain using sodium cyanide-^{13}C.[2]

References

[1]F. J. Marshall, R. E. McMahon, and W. B. Lacefield, "The Use of Diborane and Deuterodiborane in the Synthesis of Isotopically Labeled Amines," *J. Labelled Compds.*, 8, 461-473 (1972).

[2]I. Midgley, R. W. Pryor, and D. R. Hawkins, "Synthesis of [$^{13}C_2$]-Amitriptyline, Nortriptyline and Desmethylnortriptyline," *J. Labelled Compds.*, 15, 511-521 (1978).

6

Heterocyclic Compounds

The myriad procedures that have been developed for synthesis of the wide diversity of heterocyclic compounds usually have one thing in common—the starting material is a carbonyl compound and/or carboxylic acid derivative. Multiple labeling of these materials is quite common, owing to the variety of elements that can be present and the nature of the reactions, which frequently involve condensation of one relatively simple structure with another. The choice of a synthetic procedure for incorporation of the isotope into a heterocyclic ring depends to a large extent on what labeled carbonyl and carboxyl derivatives are available (or can most readily be prepared) as precursors.

3-ACETYL-4-(HYDROXYMETHYL)PYRROLE-2,3-$^{13}C_2$

($^{13}C_2$-VERRUCARIN E)

$$PhCH_2OCH_2CH=*CHCOCH_3 + 4\text{-}MeC_6H_4SO_2CH_2N*C$$

1. NaH, DMSO, Et$_2$O
2. H$_2$, Pd/C, HOAc

Procedure[1]

A solution of p-toluenesulfonylmethylisocyanide-^{13}C (1.72 g, 8.8 mmol) and 5-(benzyloxy)-3-penten-2-one-3-^{13}C (1.67 g, 8.8 mmol) in dimethylsulfoxide-ether (1:2, 43 ml) is added dropwise under nitrogen to a stirred suspension of sodium hydride (0.29 g, 12 mmol) in ether (17 ml). After 20 min the mixture is diluted carefully with water (17 ml) and extracted with ether (3 × 100 ml). The ether phase

159

is shaken with sodium chloride solution, dried over sodium sulfate, and evaporated to give the crude 3-acetyl-4-(benzyloxymethyl)pyrrole-2,3-$^{13}C_2$, which is purified by chromatography (silica gel, ether); mp 101-102° (1.62 g, 83% yield) (Note 1). The material is added to a hydrogen-saturated suspension of palladium-on-carbon (10%, 0.22 g) in acetic acid (14 ml), and the mixture is hydrogenated at room temperature and pressure until the theoretical amount of hydrogen is absorbed. The filtered solution is added slowly with stirring to ice-cooled sodium hydroxide (4 N, 60 ml). The weakly alkaline solution is extracted with methylene chloride (25 ml) (Note 2). The latter extract is filtered through silica gel and evaporated at reduced pressure, and the residue is recrystallized from benzene to afford the product, mp 91° (83% yield) (Note 1).

Notes

1 IR, UV, ^{1}H-NMR, and ^{13}C-NMR data are given.
2 From the methylene chloride extract, by preparative TLC (silica gel, methylene chloride-methanol, 98:2), is isolated starting material (0.10 g), traces of product, and 3-acetyl-4-methylpyrrole-2,3-$^{13}C_2$ (12 mg).

References

[1] A. Gossauer and K. Suhl, "Total Synthesis of Verrucarin E. Its Application to Preparation of a ^{13}C-Labeled Derivative," *Helv. Chim. Acta*, 59, 1698-1704 (1976).

5,5-DIPHENYLHYDANTOIN-2,4,5-$^{13}C_3$

$$Ph*CH(OH)*COPh + H_2N*CONH_2 \xrightarrow[NaOH]{KBrO_3}$$

Procedure[1]

To a solution of sodium hydroxide (7.6 g, 0.19 mol) and potassium bromate (1.7 g, 10 mmol) in water (14 ml) at 80° is added urea-^{13}C (3.4 g, 60 mmol) in one portion and benzoin-α,β-$^{13}C_2$ (6.0 g, 28 mmol) in several portions over a 0.5-hr period. The temperature is raised to 100°, and nitrogen is bubbled through the solution for 6 hr with replacement of lost solvent as necessary. The light-yellow mixture is diluted with water (140 ml), filtered, acidified with hydrochloric acid (6 N), and the precipitated product is collected and crystallized from ethanol-water; mp 301° (sealed tube) (4.8 g, 50% yield, Note 1).

Notes

1 The yield is based on total carbon-13: 57 mmoles in the 19 mmoles of product

and 116 mmoles (2 X 28 + 60) in the starting materials. The usual (chemical) yield, based on the limiting reagent, would be 68% (19/28), which is not as meaningful as the isotopic yield when two labeled reactants are used in non-stoichiometric amounts.

References

[1]J. A. Kepler, J. W. Lytle, and G. F. Taylor, "Synthesis of 5,5-Diphenylhydantoin-2,4,5-^{13}C$_3$," *J. Labelled Compds.,* **10**, 683-687 (1974).

DL-HISTIDINE-α-^{13}C-2,5-d_2-3-^{15}N

Procedure[1]

Ethyl 2-mercapto-4-imidazolecarboxylate-5-d-3-^{15}N   To a stirred mixture of ethyl *N*-acetylglycinate-^{15}N (15.9 g, 0.109 mol), methyl formate-d (20.0 g, 0.33 mol), and benzene (5 ml) at 6° is added a slurry of sodium methoxide (6.6 g, 0.12 mol) in benzene (25 ml). The addition is made in six portions over 30 min with maintenance of the temperature below 13°. The yellow-orange enolate mixture is refrigerated for 18 hr and then treated with ice water (55 ml) to dissolve all solid material. The aqueous phase is separated, chilled in an ice bath, and treated with hydrochloric acid (12 N, 20.5 ml, 0.25 mol) followed by potassium thiocyanate (11.5 g, 0.12 mol). The solution is heated 2 hr on a steam bath, cooled to room temperature, chilled for 4 hr, and filtered. The precipitate is collected, washed with water (10 ml), and dried to give yellow crystals, mp 182-183° (5.0 g). The filtrate is adjusted to pH 3 and chilled for 20 hr to afford additional product (total yield 6.7 g, 35%).

Ethyl 4-imidazolecarboxylate-2,5-d_2-3-^{15}N A suspension of the mercapto ester above (6.7 g, 38.5 mmol) in water-d_2 (250 ml) is warmed on a steam bath 30 min until solution is obtained, kept at room temperature 15 hr, cooled, and treated with deuterium peroxide (30%, 15 ml, 0.125 mol) over a 5-min period. The mixture is stirred for 70 hr, adjusted to pH 8 with solid sodium carbonate, and extracted with dichloromethane (8 × 50 ml). The extract is dried over magnesium sulfate and evaporated to give white crystals (3.0 g, 55% yield, Note 1).

4-(Hydroxymethyl)imidazole-2,5-d_2-3-^{15}N The ester above (2.90 g, 20 mmol) is added over 15 min to a stirred, ice-cooled mixture of lithium aluminum hydride (1.20 g, 32 mmol) and ether (35 ml). The mixture is stirred 20 hr at ambient temperature, cooled in ice, diluted with tetrahydrofuran (5 ml), and treated successively with ice water (1.2 ml), sodium hydroxide (15%, 1.2 ml), and water (3.6 ml). The solid from filtration is washed with tetrahydrofuran, suspended in hot methanol, saturated with carbon dioxide, and filtered. The process is repeated twice, and the combined methanol filtrates are evaporated. A solution of the residue in water (40 ml) is adjusted to pH 6-7 with acetic acid and treated with a solution of picric acid (4.3 g, 19 mmol) in warm water (100 ml) to give a yellow precipitate, mp 200-202° (4.30 g). The picrate, Dowex-2(Cl⁻) resin (20 g), and water (100 ml) are warmed on a steam bath for 15 min, stirred 20 hr, filtered, and evaporated at reduced pressure to afford the product as the crystalline hydrochloride hydrate (2.1 g, 67% yield).

DL-Histidine-α-^{13}C-2,5-d_2-3-^{15}N The above-mentioned product (2.1 g, 13.5 mmol) and thionyl chloride (5 ml) are refluxed with stirring for 45 min, evaporated, and twice treated with carbon tetrachloride (20 ml) followed by evaporation to give crude 4-(chloromethyl)imidazole-2,5-d_2-3-^{15}N hydrochloride (2.2 g). A solution of sodium ethoxide [from sodium (0.62 g, 26.8 mmol) and ethanol (23 ml)] is treated with ethyl acetamidocyanoacetate-2-^{13}C (2.29 g, 13.4 mmol), chilled to 0-5°, and rapidly added to the chloromethyl compound. The mixture is stirred at room temperature for 15 hr, the solvent is evaporated at reduced pressure, and the residue is recrystallized from water (7 ml) to afford N-acetyl-α-cyanohistidine-α-^{13}C-2,5-d_2-3-^{15}N ethyl ester as white crystals, mp 85° (2.63 g, 77% yield, Note 2).

A solution of the cyano ester (2.06 g, 8.1 mmol) in sulfuric acid (3 N, 16 ml) is heated 18 hr at 100°, cooled to room temperature, adjusted to pH 8-9 with barium hydroxide (saturated), filtered through Celite, and evaporated at reduced pressure. A solution of the residue in water (15 ml) is saturated with carbon dioxide, filtered, and evaporated at reduced pressure. The residue is recrystallized from ethanol (10 ml) to afford the product as white crystals (1.00 g, 78% yield, Note 3).

Notes

1   Unlabeled product has been obtained in 72% yield, mp 155-157°. Sometimes the ester partially hydrolyzes to the acid, which is reconverted to the ester by evaporation of the aqueous mother liquor and treatment with ethanolic hydrogen chloride.

2 The melting point of purified material is 100-103°; however, the product is sufficiently pure for the next step.

3 The product is shown to be pure by paper chromatography, ^{1}H-NMR and ^{13}C-NMR, and GC/Ms.

Other Preparations

L-Histidine-2-^{13}C has been prepared[2] on a 0.3-mol scale using a procedure based on that for the carbon-14 compound.[3] Potassium thiocyanate-^{13}C (from silver cyanide-^{13}C in quantitative yield) and 2,5-diaminolevulinic acid give 2-mercaptohistidine-2-^{13}C (69% yield), which reacts with ferric sulfate to afford the product in 55% overall yield following purification through the flavianate.

References

[1] C. SooHoo, J. A. Lawson, and J. I. DeGraw, "Synthesis of Multilabeled Histidine," *J. Labelled Compds.*, **13**, 97-102 (1977).

[2] A. Murray, in "Annual Report of the Biological and Medical Research Group (H-4)," C. R. Richmond and G. L. Voelz, Eds., Los Alamos Scientific Laboratory Report LA-4923-PR, April 1972, p. 122.

[3] A. Murray and D. L. Williams, *Organic Syntheses with Isotopes,* Interscience, New York, 1958, pp. 294-296.

CREATININE-2-^{15}N

[2-(AMINO-^{15}N)-1-METHYL-2-IMIDAZOLIN-4-ONE]

Procedure[1]

Ammonia-^{15}N (144 mg, 8 mmol) is condensed in a solution of 1-methyl-2-(methylthio)-4-oxoimidazolium iodide (544 mg, 2 mmol) in ethanol (4 ml) contained in a pressure tube (2.8 × 10 cm), and the mixture is held at 20° for 12 hr. The crystallized product is filtered, washed with ethanol, and dried. A second crop is obtained from the mother liquor (total yield 181 mg, 80%); mp 258°.

References

[1] R. Medina and H.-L. Schmidt, "Specific ^{15}N-Labeling of Creatinine," *J. Labelled Compds.*, **12**, 565-569 (1976).

$^{15}N_1$-2-(PHENYLAMINO)-2-THIAZOLINE

Method I

$$Ph*NCS + HOCH_2CH_2{}^\bullet NH_2 \xrightarrow{Et_2O} Ph*NHCS^\bullet NHCH_2CH_2OH$$

$$\xrightarrow[\text{2. NH}_4\text{OH}]{\text{1. HCl, }100^\circ} \quad \text{[thiazoline ring]}{-}*NHPh$$

Procedure[1]

2-(Phenylamino-^{15}N)-2-thiazoline   A solution of phenylisothiocyanate-^{15}N (0.68 g, 5 mmol) in ether (5 ml) is treated with ethanolamine (0.305 g, 5 mmol). After a day the precipitated colorless crystals of 1-(2-hydroxyethyl)-3-phenylthiourea-3-^{15}N are filtered and washed with ether; mp 137-138° (0.98 g, 100% yield). A solution of the product in hydrochloric acid (3 ml) is heated at 100° for 1 hr, cooled to 0°, and brought to pH 9 with ammonium hydroxide. The colorless crystals are filtered and washed with water to afford the product (0.797 g), which is recrystallized from methanol (8 ml); mp 159-160° (0.577 g, 64% yield).

2-(Phenylamino)-2-thiazoline-3-^{15}N   In the same manner, from ethanolamine-^{15}N hydrochloride (0.5 g, 5 mmol) in methanolic potassium hydroxide is obtained 1-(2-hydroxyethyl)-3-phenylthiourea-1-^{15}N (0.8 g, 80% yield), which is cyclized to the product, mp 158-159° (0.613 g, 85% yield).

Method II

$$Ph*NH_2 + \text{[ring]}{-}SCH_3 \cdot HI \xrightarrow[100^\circ]{H_2O} \text{[ring]}{-}*NHPh$$

Procedure[1]

2-(Phenylamino-^{15}N)-2-thiazoline   To a solution of 2-(methylthio)-2-thiazoline hydroiodide (1.1 g, 4.2 mmol) in water (3 ml) is added aniline-^{15}N (0.282 g, 3 mmol), and the solution is heated 3 hr at 100°, cooled, filtered, and made alkaline. The precipitated crystals are recrystallized from methanol to give the product, mp 160° (0.48 g, 90% yield).

Other Preparations

By Method I, 2-(phenylimino-^{15}N)perhydrothiazine is prepared from phenyliso-thiocyanate-^{15}N and 3-aminopropanol (55% yield). 2-(Phenylimino)perhydrothiazine-3-^{15}N is similarly obtained from phenylisothiocyanate and 3-amino-propanol-^{15}N (54% yield).[1]

References

[1] J. Volford and D. Banfi, "Synthesis of ^{15}N-Labeled 2-Substituted 2-Thiazolines and Analogous Thiazines," *J. Labelled Compds.*, **11**, 419-426 (1975).

THIOPHENE-2-^{13}C OR THIOPHENE-2,5-^{13}C$_2$

$$\text{NaOO*CCH}_2\text{CH}_2\text{*COONa} \xrightarrow[200-300°]{P_4S_7} \text{[thiophene ring]}$$

Procedure[1]

Thiophene-2-^{13}C A sealed tube containing sodium cyanide (3.30 g, 66 mmol), sodium 3-chloropropionate (8.74 g, 66 mmol), ethanol (40 ml), and water (20 ml) is heated with shaking for 5 hr at 110°. The mixture is dissolved in water-ethanol (1:6, 150 ml) containing sodium hydroxide (22.5 g, 0.56 mol), and the solution is refluxed for 3 hr and cooled. The precipitated disodium succinate-1-^{13}C is filtered, washed with ethanol, dried at 100°, mixed carefully with phosphorus heptasulfide (34 g, 97.8 mmol), and transferred to the pyrolysis apparatus (Note 1). The mixture is heated under nitrogen to 200°, and over 1.5 hr the temperature is raised to 300° to afford the pure product (1.10 g, 19% yield).

Thiophene-2,5-^{13}C$_2$ A sealed tube containing sodium cyanide-^{13}C (6.66 g, 134 mmol), 1,2-dibromoethane (12.6 g, 67 mmol), ethanol (15 ml), and water (15 ml) is heated 3.5 hr at 100°. The solvent is removed, chloroform (200 ml) is added, and the mixture is stirred for 15 min. The residue after evaporation of the chloroform is hydrolyzed as described above to give disodium succinate-1,4-^{13}C$_2$, which is converted by the procedure above to the product (1.10 g, 19% yield).

Notes

1 A sketch is given by de Jong, Sinnige, and Janssen of the apparatus, which is constructed of glass with a thermometer well into the reaction mixture, a cold-finger condenser from which the condensate of product drips into an outlet, and nitrogen inlet-outlet.

References

[1] F. de Jong, H. J. M. Sinnige, and M. J. Janssen, "Carbon Skeletal Rearrangement and Hydrogen Migration in Thiophene," *Org. Mass Spectrom.*, **3**, 1539-1549 (1970).

$^{13}C_1$-1,2,3-THIADIAZOLE

Method I

$$CH_3*COCOOH \xrightarrow[PhCH_3, Et_2O]{H_2NNHCOOEt} CH_3*C(COOH)=NNHCOOEt$$

$$\xrightarrow{SOCl_2} HOOC\text{—}\underset{S}{\overset{*}{\big|}}\overset{N}{\underset{N}{\big\|}} \xrightarrow{260°} \underset{S}{\overset{*}{\big|}}\overset{N}{\underset{N}{\big\|}}$$

Procedure[1]

1,2,3-Thiadiazole-4-^{13}C Sodium pyruvate-2-^{13}C (1.0 g, 9 mmol) is stirred with hydrochloric acid (conc., 2 ml), the solution is extracted with ether (15 ml), and the extract containing the pyruvic-2-^{13}C acid is treated with carboethoxyhydrazine (1.2 g, 13 mmol) in warm toluene (3 ml). The mixture is stirred for 2.5 hr, most of the solvent is decanted from the oil, and the remainder is removed under vacuum. Thionyl chloride (2 ml) is added to the residue, and the mixture is stirred for 4 hr. The precipitate is filtered and washed with ethyl acetate to give 1,2,3-thiadiazole-4-carboxylic-4-^{13}C acid (0.53 g, 44% yield). The acid is immersed in a Wood's metal bath preheated to 260°. After the vigorous reaction the liquid formed is extracted into chloroform-d (0.5 ml) and purified by vapor phase chromatography (SE 30 column, 90°) to afford the product (74 mg, 58% yield) (Note 1).

Method II

$$*CH_3CN \xrightarrow[\substack{2.\ H_2O \\ 3.\ H_2NNHCOOEt}]{\substack{1.\ (i\text{-}Bu)_2AlH, \\ PhMe,\ n\text{-}C_7H_{16}}} *CH_3C=NNHCOOEt \xrightarrow{SOCl_2} \underset{S}{\overset{*}{\big|}}\overset{N}{\underset{N}{\big\|}}$$

Procedure[1]

1,2,3-Thiadiazole-5-^{13}C Acetonitrile-2-^{13}C (0.5 g, 12 mmol) in toluene (5 ml) is reduced under rigorously anhydrous conditions with diisobutylaluminum hydride in heptane (1 M, 13 ml), which is added at ice-bath temperature over 6 min. The reaction is stirred an additional 30 min and treated with water (13 ml). A solution of ethoxycarbonylhydrazine (1.27 g, 12 mmol) in toluene (3 ml) is immediately added. After 5 min chloroform (10 ml) is added and the precipitate is filtered and washed with chloroform (200 ml). The chloroform extract is evaporated at reduced pressure to give the crude hydrazone (0.86 g, 55% yield). Thionyl chloride (4 ml) is added, and the mixture is refluxed for 1 hr, hydrolyzed, neutralized with sodium bicarbonate, and extracted with ether (250 ml). The extract, dried over magnesium sulfate and concentrated by distillation at atmospheric pressure, gives the product (257 mg, 45% yield), which is purified by chromatography (Carbowax-20M, 90°) (Notes 2 and 3).

Notes

1 Decarboxylation of the carboxylic acid intermediate, after isotopic exchange reaction with water-d_2, gives 1,2,3-thiadiazole-4-^{13}C-4-d. Base-catalyzed isotopic exchange of 1,2,3-thiadiazole-4-^{13}C with sodium deuteroxide in water-d_2 gives 1,2,3-thiadiazole-4-^{13}C-5-d.

2 Isotopic exchange of the product with sodium deuteroxide and water-d_2 gives 1,2,3-thiadiazole-5-^{13}C-5-d.

3 IR and NMR spectral data are given for all products.

References

[1] A. Krautz and J. Laureni, "Synthesis of ^{13}C-Labelled and Doubly Labelled (^{13}C, ^{2}H) 1,2,3-Thiadiazoles. Precursors to Isotopically Labelled Thiirenes," *J. Labelled Compds.*, 15, 697-702 (1978).

PYRIDINE-^{15}N

Procedure[1]

To a refluxing solution of ammonium-^{15}N chloride (22.7 g, 0.42 mol), methylene blue (157 g, 0.42 mol), and sulfuric acid (conc., 50 ml, 0.90 mol) in water (1.7 liter) is added 2-ethoxy-3,4-dihydro-2H-pyran (53.8 g, 0.42 mol) dropwise over a period of 2 hr. The mixture is refluxed an additional 10 hr. The reflux condenser and dropping funnel are replaced with a distilling head, water (500 ml) is added, and the mixture is distilled until the odor of glutaraldehyde is no longer detected in the distillate (approx. 1.2 liter). The reaction mixture is cooled, made basic by gradual addition of sodium hydroxide (50%, 200 ml), and distilled until approx. 500 ml of distillate is collected. Redistillation of the distillate through a 6-cm column packed with glass helices gives a solution (bp to 93° at 580 mm Hg, 115 g) that contains the product (20.7 g, 62% yield by UV analysis). The mixture is acidified with hydrochloric acid, evaporated to a semisolid mass, and made basic with the minimum amount of sodium hydroxide (50%). Molecular sieve (Type 3A, 300 g) is added slowly to the chilled mixture, and the product is collected by bulb-to-bulb vacuum transfer. A second transfer from molecular sieve (10 g) removes a trace of water to give the anhydrous product (18.2 g, 55% yield).

References

[1] T. W. Whaley and D. G. Ott, "Syntheses with Stable Isotopes: Pyridine-^{15}N," *J. Labelled Compds.*, 10, 283-286 (1974).

NICOTINIC-^{13}C ACID

$$Na*CN \xrightarrow[\text{NaHSO}_3]{\text{CuSO}_4} Cu*CN$$

Procedure[1]

Cuprous cyanide-[13]C A solution of sodium bisulfite (3.66 g, 26.5 mmol) in water (10 ml) at 50-60° is slowly added with stirring to a solution of cupric sulfate (13.2 g, 52.6 mmol) in water (42 ml) at 40-50°. The resulting solution is added immediately and quickly to a solution of sodium cyanide-[13]C (1.78 g, 36.3 mmol) in water (7 ml) at 50-80°, and the mixture is allowed to stand for 10 min. The precipitate is filtered, washed with boiling water, cold ethanol, and dried at 120° overnight to give a beige powder (approx. quantitative yield).

Nicotinic-^{13}C acid In a 100-ml flask fitted with a reflux condenser and magnetic stirrer, cuprous cyanide (2.7 g, 30 mmol) and 3-bromopyridine (4.4 ml, 45.6 mmol) are heated by an oil bath at 165-170° for 1 hr (Note 1). The solidified tar is resuspended in ethanol (40 ml) containing sodium hydroxide (4 g, 0.1 mol). The mixture is refluxed for 3 hr, evaporated to dryness, redissolved in sodium hydroxide (2 N, 80 ml), filtered, concentrated, and applied to a Dowex-1-X8 column (formate form, 50-100 mesh, 20 × 2.1 cm). The column is eluted with water (120 ml) and formic acid (0.25 N, 500 ml). The ultraviolet-absorbing fractions are combined, concentrated, applied to a Chelex column (50-100 mesh, 20 × 2.1 cm), and eluted with water (Note 2). The ultraviolet-absorbing fractions are combined, concentrated, and applied to a Dowex-1-X8 column (formate, 100-200 mesh, 22 × 2.0 cm). The column is washed with water (100 ml), and the product is eluted with formic acid (0.25 N, 150 ml). The combined ultraviolet-absorbing fractions are evaporated to give a white powder, mp 235° (3.2 g, 72% yield from sodium cyanide) (Note 3).

Notes

1 The black viscous solution may begin to solidify by the end of the heating.
2 The column removes residual copper ions.
3 The yield is about three times that previously reported for this method and is attributed to hydrolysis of the intermediate nitrile without isolation.

References

[1]M. C. Meinert, H. A. Nunez, and R. U. Byerrum, "The Synthesis of Nicotinic Acid-7-^{13}C," *J. Labelled Compds.*, **14**, 893-896 (1978).

NICOTINAMIDE-2-^{13}C

$$N^*CCH_2COOMe + (MeO)_2CHCH_2CH(MeO)_2 \xrightarrow[Ac_2O]{ZnCl_2}$$

$$MeOCH=CHCH=C(^*CN)COOMe \xrightarrow[HOAc]{HBr}$$

$$\xrightarrow[Pd/BaCO_3]{H_2}$$

$$\xrightarrow[H_2O]{NH_3}$$

Procedure[1]

Methyl 2-(cyano-^{13}C)-5-methoxy-2,5-pentadienoate A mixture of 1,1,3,3-tetra-methoxypropane (12.3 g, 7.5 mmol), acetic anhydride (25 ml), and zinc chloride (68 mg) is heated under reflux, and methyl (cyano-^{13}C)acetate (4.95 g, 5.0 mmol) is added dropwise. Reflux is maintained for 18 hr, and the volatile materials are distilled until the distillate temperature reaches 122°. The residue is cooled and filtered. The filtrate, which solidifies on standing, is distilled (Kugel-Rohr) to afford a yellow oil, bp 90°, and the product as a yellow solid, bp 110-140° with partial decomposition (7.5 g, 90% yield).

Methyl 2-bromonicotinate-2-^{13}C The above-mentioned vinyl ether (1.20 g, 7.1 mmol) in acetic acid (9 ml) at 40° is treated dropwise with an acetic acid solution (18 ml) that had been saturated with hydrogen bromide at 0° while the temperature is maintained at 40-45°. The temperature is raised to 55° for 30 min, and the darkened solution is cooled, poured into water, neutralized by careful addition of sodium carbonate, and extracted with methylene chloride (3 × 225 ml). The extract is washed with water, dried over sodium sulfate, and evaporated, and the residue is distilled (Kugel-Rohr) to afford the product, bp 105-120° at 35 mm Hg (1.46 g, 97% yield).

Methyl nicotinate-2-^{13}C Methyl 2-bromonicotinate-2-^{13}C (1.65 g, 7.6 mmol) is added at room temperature to a vigorously stirred suspension of palladium-on-barium carbonate (1%, 10 g) in ethanol (150 ml) under hydrogen at atmospheric pressure. After the hydrogen (171 ml) is absorbed the solution is filtered and evaporated at reduced pressure, and the residual oil is distilled to give the product, bp 110-125° at approx. 25 mm Hg, mp 42-43° (0.98 g, 93% yield).

Nicotinamide-2-^{13}C Ammonia is bubbled through a solution of methyl nicotinate (0.98 g, 7.1 mmol) in water (150 ml) at 3° for 6 hr and then at room temperature for 12 hr. The solution is extracted with ether (50 ml), and the extract is washed with water (25 ml) and lyophilized to afford a white solid, mp 130-131° (0.65 g, 75% yield).

Other Preparations

Nicotinamide-6-^{13}C has been prepared from 2-(tribromomethyl)quinoline-2-^{13}C by way of dimethyl 2,5-pyridinedicarboxylate-2-^{13}C, 5-methoxycarbonyl-2-pyridine-carboxylic-2^{13}C acid, and methyl nicotinate-6-^{13}C.[2] A procedure has been proposed for preparing nicotinamide-4-^{13}C and nicotinamide-5-^{13}C.[3]

References

[1]T. A. Bryson, J. C. Wisowaty, R. B. Dunlap, R. R. Fisher, and P. D. Ellis, "Biological Probes. II. Ring Labeled Nicotinamide," *J. Org. Chem.*, 39, 3436-3438 (1974).

[2]T. A. Bryson, J. C. Wisowaty, R. B. Dunlap, R. R. Fisher, and P. D. Ellis, "Biological Probes. I. Carbon-6-Labeled Nicotinamide," *J. Org. Chem.*, 39, 1158-1160 (1974).

[3]T. A. Bryson, D. M. Donelson, R. B. Dunlap, R. R. Fisher, and P. D. Ellis, "Biological Probes. 3. Methods for Carbon-4 and Carbon-5 Labeling in Nicotinamide," *J. Org. Chem.*, 41, 2066-2067 (1976).

NICOTINAMIDE-1-^{15}N

Procedure[1] (Note 1)

Ammonium-^{15}N chloride (0.93 g, 17 mmol) and triethylamine (3.35 g, 33.2 mmol) in methanol (600 ml) are stirred for 20 min, and a solution of 3-carbamoyl-1-(2,4-dinitrophenyl)pyridinium chloride (5.28 g, 16.3 mmol) in methanol (100 ml) is added over 1 hr (Note 2). The solution is stirred at room temperature for 3 days (Note 3), the methanol is evaporated at reduced pressure, and the yellow precipitate is suspended in water (100 ml) and filtered. The yellow solid (dinitroaniline) is washed with cold water (40 ml total), and the combined filtrates are concentrated at reduced pressure. The dried residue is dissolved in acetonitrile (80 ml), which gives on standing long needles of triethylamine hydrochloride. The filtered solution is evaporated, dissolved in minimal water, and applied to a reverse-phase chromatography column (C-2 silica gel, 2.5 × 40 cm). The column is eluted with water (100 ml) and then with water-ethanol linear gradient (1.5 liter). The product fractions are evaporated to a pale-yellow powder, mp 118-123°, and the residual dinitroaniline is separated by column chromatography (silica gel, 2 × 30 cm, acetonitrile-ethanol-dichloromethane, 5:1:1). The product fractions are evaporated to dryness, and the residue is recrystallized from acetonitrile to give the product, mp 128-129° (1.65 g, 83% yield).

Notes

1 This procedure represents a specific example of a potentially general and facile method for the synthesis of a wide variety of nitrogen-15 labeled pyridine heterocycles. A mechanism for reaction of the Zincke salt with ammonia is shown.[1]

2 A red color, characteristic of the ring-opened form of the intermediate, is produced immediately.

3 The solution is stirred until the color has become yellow.

References

[1] N. J. Oppenheimer, T. O. Matsunaga, and B. L. Kam, "Synthesis of ^{15}N-1 Nicotinamide. A General, One Step Synthesis of ^{15}N Labeled Pyridine Heterocycles," *J. Labelled Compds.*, 15, 191-196 (1978).

DL-CONIINE-^{15}N
(2-PROPYLPIPERIDINE-^{15}N HYDROCHLORIDE)

$$*NH_4Br + CH_3(CH_2)_2CO(CH_2)_3CHO \xrightarrow[\text{2. HCl}]{\text{1. NaCNBH}_3, \text{ MeOH}}$$

Procedure[1]

A solution of 5-oxooctanal (Note 1) in methanol (10 ml) containing sodium cyanoborohydride (100 mg, 1.6 mmol) and ammonium-^{15}N bromide (99 mg, 1 mmol) is stirred for 3 days, acidified with hydrochloric acid, and evaporated to dryness. The residue is made basic with sodium hydroxide and extracted with ether (4 × 10 ml). The extract is dried over magnesium sulfate and evaporated at 10°. The residue is distilled at 120° and 0.1 mm Hg, and the product is collected in a U-tube cooled in liquid nitrogen, dissolved in ether, and treated with ethanolic hydrogen chloride to afford colorless needles, mp 215-216° (96 mg, 58% yield).

Notes

1 The keto aldehyde is obtained from 1-propyl-1,2-cyclopentanediol (144 mg, 1.00 mmol) by periodate oxidation.

References

[1] E. Leete, H. V. Isaacson, and H. D. Durst, "Synthesis of ^{15}N Labelled Alkaloids: Coniine-^{15}N and Nicotine-1'-^{15}N," *J. Labelled Compds.*, 7, 313-317 (1971).

TRYPTOPHAN-α-^{15}N

Procedure[1]

To a solution of 3-indolylpyruvic acid (205 mg, 1.0 mmol) and ammonium-^{15}N nitrate (95 mg, 1.2 mmol) in methanol (15 ml) is added sodium cyanoborohydride (190 mg, 3 mmol). The solution is adjusted to pH 7 (Note 1), stirred at 25° for 36 hr, and evaporated. The residue is dissolved in water (3 ml) and applied to an ion exchange column (Dowex-50(H$^+$), 100 meq). The column is washed with water (500 ml) and eluted with ammonium hydroxide (1 N, 300 ml) to afford the product as an amorphous powder (47 mg, 23% yield).

Notes

1   No color change is given by red or blue litmus. Adjustment of the pH is important; at pH 9 only polymeric material is obtained.

Other Preparations

The preparation of other amino acids by reductive amination of substituted pyruvic acids has also been studied and appears well-suited for nitrogen-15 labeling.

References

[1] R. F. Borch, M. D. Bernstein, and H. D. Durst, "The Cyanohydroborate Anion as a Selective Reducing Agent," *J. Am. Chem. Soc.*, 93, 2897-2904 (1971).

DL-NICOTINE-1'-^{15}N

Method I

Procedure[1]

DL-Nornicotine-1'-¹⁵N A mixture of formamide-¹⁵N (2.99 g, 88.1 mmol), cyclopropyl 3-pyridyl ketone (4.36 g, 30 mmol), and magnesium chloride hexahydrate (1.19 g, 5.9 mmol) in 2-ethoxyethyl ether (15 ml) is refluxed under nitrogen with stirring for 21 hr, cooled, and treated with hydrochloric acid (conc., 20 ml) at a temperature less than 20°. The solution is extracted with chloroform (3 × 40 ml), the chloroform extract is washed with hydrochloric acid (10%, 10 ml), and the combined aqueous acid layers are heated at reflux under nitrogen for 16 hr. The solution is cooled, made basic (pH 11) with sodium hydroxide (50%), and extracted with ether (4 × 40 ml). The extract is dried over sodium hydroxide and evaporated to give a crude oil (2.79 g). Continuous extraction of the aqueous layer and insoluble solids gives additional oil (0.26 g). The crude oil is distilled to afford the product, bp 77.5-81° at 0.2 mm Hg (2.36 g, 19% yield) (Note 1).

DL-Nicotine-1'-¹⁵N DL-Nornicotine-1'-¹⁵N (1.36 g, 9.2 mmol) is added with the aid of water (4 × 0.5 ml) to a stirred solution of formic acid (88%, 2.40 g, 4.6 mmol) and formaldehyde solution (37%, 1.36 g, 9.2 mmol) at 4°. The mixture is refluxed under nitrogen with stirring for 5 hr, cooled to room temperature, left overnight, cooled in an ice bath, made basic with sodium hydroxide (50%), and extracted with ether (4 × 20 ml). The extract is dried over sodium hydroxide and evaporated at reduced pressure to give the crude product as an oil (1.29 g). The aqueous layer is continuously extracted with ether for 16 hr to afford an additional amount (0.08 g). The crude oil is distilled at an air-bath temperature of 93-119° and 0.37 mm Hg to afford the product (1.25 g, 84% yield).

Method II

Procedure[2]

DL-Nornicotine-1'-¹⁵N 4,4-(Ethylenedioxy)-4-(3-pyridyl)butanal (761 mg, 3.68 mmol) in water (2 ml) and methanol saturated with hydrogen chloride (4 ml) is stirred at room temperature for 3 hr. The solvent is removed, and the residue is treated with a solution of ammonium-¹⁵N bromide (588 mg, 7.36 mmol) and sodium cyanoborohydride (243 mg, 3.81 mmol) in methanol (30 ml). The mixture is stirred at room temperature for 2 days, adjusted to pH 2 with hydrochloric

acid, stirred 3 hr, and evaporated. The residue is made strongly alkaline with potassium hydroxide (Note 2) and extracted with ether in a continuous extractor for
2 days. The extract is dried over magnesium sulfate and evaporated to afford the
crude product (309 mg, 28% yield) (Note 2).

DL-Nicotine-1'-^{15}N diperchlorate The crude nornicotine above (2 mmol) is added
to a mixture of formic acid (97%, 443 mg) and formaldehyde solution (40%, 1 ml)
at 0°. After 1 hr the mixture is heated on a steam bath for 16 hr, cooled, made
strongly alkaline with potassium hydroxide, and extracted with ether for 24 hr. The
extract is dried over magnesium sulfate and evaporated. The residual oil is dissolved
in methanol (5 ml), cooled, and treated with perchloric acid (70%, 0.25 ml). Ethyl
acetate is added to precipitate the product as colorless plates, mp 206-208° (567
mg, 78% yield).

Notes

1 Various ratios of reactants and reaction conditions have been studied.
2 Recovery of ammonia-^{15}N is apparently not carried out; presumably the isotopic
   yield could be doubled.

References

[1]W. B. Edwards III, D. F. Glenn, F. Green, and R. H. Newman, "The Preparation of Tobacco
Constituents Incorporating Stable Isotopes, I. The Synthesis of d,1-Nornicotine-1'-^{15}N and
d,1-Nicotine-1'-^{15}N," *J. Labelled Compds.*, **14**, 255-261 (1978).
[2]E. Leete, H. V. Isaacson, and H. D. Durst, "Synthesis of ^{15}N Labelled Alkaloids: Coniine-^{15}N
and Nicotine-1'-^{15}N," *J. Labelled Compds.*, **7**, 313-317 (1971).

4-ETHOXY-3,6-DIPHENYLPYRIDAZINE-5-^{13}C

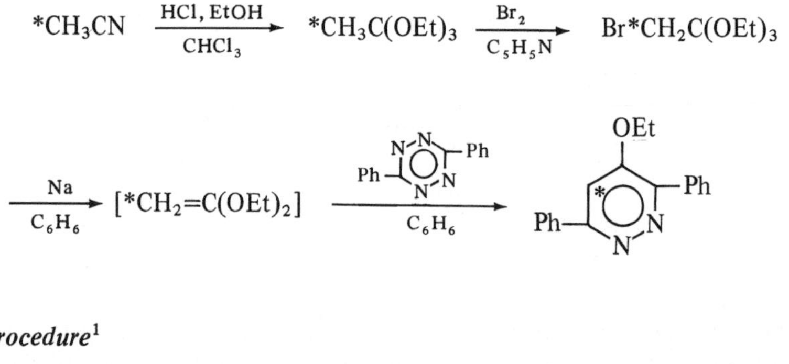

Procedure[1]

1,1,1-Triethoxyethane-2-^{13}C (triethyl orthoacetate-1-^{13}C) Hydrogen chloride
(9.17 g, 0.252 mol) is added to a water-cooled mixture of acetonitrile-2-^{13}C (10.3

g, 0.252 mol), ethanol (11.6 g, 0.252 mol), and chloroform (9ml). The temperature is allowed to rise to room temperature, and the mixture is agitated occasionally over a period of 48 hr. Ethanol (50 ml) is added, and, after 2 days, the ammonium chloride is filtered and washed with ethanol. The filtrate is treated with sodium hydroxide (5%, 200 ml), and the product is dried (potassium carbonate), concentrated, and distilled to give the product, bp 70-80° at 60 mm Hg (11.6 g, 28% yield, Note 1).

2-Bromo-1,1,1-triethoxyethane-2-^{13}C To the above-mentioned orthoacetate (11.6 g, 72 mmol) bromine (11.5 g, 72 mmol) is added dropwise over 2 hr with the temperature held at 0-5° to afford the product, bp 88-92° at 20 mm Hg (8.7 g, 50% yield, Note 2).

4-Ethoxy-3,6-diphenylpyridazine-5-^{13}C To a stirred mixture of powdered sodium (1.7 g, 72 mmol) and benzene (36 ml) heated under gentle reflux is added the above bromo ester (8.7 g, 36 mmol) over a period of 45 min. The mixture is stirred an additional 2 hr at reflux, cooled, and the clear supernatant liquid is filtered from the precipitated blue salts, which are washed by several triturations with benzene (Note 3). Small portions of 3,6-diphenyltetrazine are added to the refluxing benzene solution over a period of approx. 2 days until the characteristic violet color of the tetrazine fails to disappear. The solvent is evaporated at reduced pressure, and the excess tetrazine is removed by several washings of the residue with petroleum ether (Note 4) to afford the product, mp 102-104° (3.0 g, 30% yield).

Notes

1 The reference on which the preparation is based gives yields of 59-78%.

2 The reference on which the preparation is based gives a yield of 74%.

3 The 1,1-diethoxyethene-2-^{13}C (ketene-2-^{13}C diethyl acetal) is not isolated; the reference preparation gives a yield of 66% of distilled product.

4 The color changes from violet to pale yellow.

Other Preparations

Acid hydrolysis of the product affords 3,6-diphenyl-4(1H)-pyridazinone-5-^{13}C (92% yield), which is converted with phosphoryl chloride to 4-chloro-3,6-diphenyl-pyridazine-5-^{13}C (85% yield). Potassium amide in liquid ammonia gives 4-amino-3,6-diphenylpyridazine-4,5-^{13}C$_1$ (21% yield), accompanied by imino-4,4′-bis(3,6-diphenylpyridazine-4,5-^{13}C$_1$).

References

[1]D. E. Klinge and H. C. van der Plas, "Didehydrohetarenes (XXXVII). On the existence of 4,5-didehydropyridazine," *Rec. Trav. Chim.*, 95, 34-36 (1976).

1,2-DIHYDRO-3,6-PYRIDAZINEDIONE-3-^{13}C (^{13}C$_1$-MALEIC HYDRAZIDE)

Procedure[1]

A mixture of maleic-1-^{13}C anhydride (1.147 g, 11.70 mmol), hydrazine sulfate (1.522 g, 11.70 mmol), and water (11.8 ml) is stirred and heated under reflux for 4 hr, cooled in ice, and filtered. The solid is washed with cold water (15 ml), ethanol (3 ml), and ether (3 ml) to give the white crystalline product, mp 306-307° (1.027 g, 78% yield) (Note 1).

Notes

1 In some experiments the product gives a positive test for hydrazine, which can be readily removed by recrystallization from water.

Other Preparations

By the same procedure are prepared 1,2-dihydro-3,6-pyridazinedione-4-^{13}C (from maleic-2-^{13}C anhydride obtained from aspartic-3-^{13}C acid) and 1,2-dihydro-3,6-pyridazinedione-^{15}N$_2$ (in 86% yield from hydrazine-N$_2$ sulfate).

References

[1] J. M. Patterson, L. L. Braun, N. F. Haidar, J. C. Huang, and W. T. Smith, Jr., "A Novel Synthesis of Maleic Anhydride-1- and -2-^{13}C and Its Subsequent Conversion to Labeled Maleic Hydrazide," J. Labelled Compds., 14, 439-443 (1978).

URACIL-6-^{13}C

$$H_2NCONHCOCH_2*CN \xrightarrow[\text{Raney-Ni, H}_2]{\text{PhNH}_2 \cdot \text{HCl, H}_2\text{O}} H_2NCONHCOCH=*CHNHPh$$

Procedure[1]

1[3-(Phenylamino)propenoyl-3-^{13}C]urea A suspension of [(cyano-^{13}C)acetyl] urea (3.60 g, 28.3 mmol) and activated Raney nickel (approx. 1.0 g) in water (125 ml) containing aniline hydrochloride (4.04 g, 31.2 mmol) is agitated at room temperature under hydrogen at 35 psi for 10 hr. The resultant paste is extracted with alcohol several times until the catalyst regains its black color, and the extracts are concentrated at reduced pressure and cooled to afford the product as a crystalline solid, mp 208° dec. (2.30 g, 40% yield).

Uracil-6-^{13}C The product above (1.5 g, 7.3 mmol) is dissolved in alcohol that has been saturated at room temperature with hydrogen chloride. The suspension is kept at 4° for 4 days and filtered. The solid is washed with cold ethanol and extracted with ammonium hydroxide. The filtered extract is adjusted to pH 3 with sulfuric acid (conc.) to give the product as a crystalline powder (0.39 g, 48% yield).

References

[1]J. W. Triplett, S. W. Mack, S. L. Smith, and G. A. Digenis, "Synthesis of Carbon-13 Labelled Uracil, 6,7-Dimethyllumazine and Lumichrome, via a Common Intermediate: Cyanoacetylurea," *J. Labelled Compds.*, **14**, 35-41, (1978).

THYMINE-2,6-^{13}C$_2$

$$CH_3CH(*CN)COOH + H_2N*CONH_2 \xrightarrow[90°]{Ac_2O} CH_3CH(*CN)CONH*CONH_2$$

Procedure[1]

1-[2-(Cyano-^{13}C)propionyl]urea-^{13}C A mixture of urea-^{13}C (6.84 g, 0.112 mol), 2-(cyano-^{13}C)propionic acid (11.21 g, 0.112 mol), and acetic anhydride (12.59 g, 0.123 mol) is heated with stirring to 90° and maintained at that temperature for 2 hr (Note 1). The cooled reaction mixture is filtered, and the product is washed with ether until the yellow color is removed to give white crystals, mp 191-192° (10.20 g, 64% yield) (Note 2).

Thymine-2,6-^{13}C$_2$ A solution of the above cyanoacylurea (2.86 g, 20.0 mmol) in hot acetic acid (40 ml) is added to a hydrogenation vessel containing prereduced Adams' catalyst (1.14 g) in water (40 ml), and the mixture is heated at 70° until hydrogen uptake ceases and filtered while hot. The filtrate is concentrated at reduced pressure until a white precipitate appears, cooled overnight, and filtered. Additional crystals are obtained by further concentration of the filtrate to afford the product (total yield 1.95 g, 59%) (Notes 2 and 3).

Notes

1 The starting material is in solution when 80° is reached; a white solid precipitates after approx. 15 min at 90°.
2 IR and ^{13}C-NMR data are given.
3 A discussion of the mechanism of α-cyanoacylurea hydrogenations and conditions leading to increased yields is given by Redwine and Whaley.

References

[1]C. M. Redwine and T. W. Whaley, "Syntheses with Stable Isotopes: Thymine-2,6-^{13}C$_2$," *J. Labelled Compds.*, **16**, 315-319 (1979).

2′,3′,5′-TRI-*O*-BENZOYLURIDINE-4-^{13}C

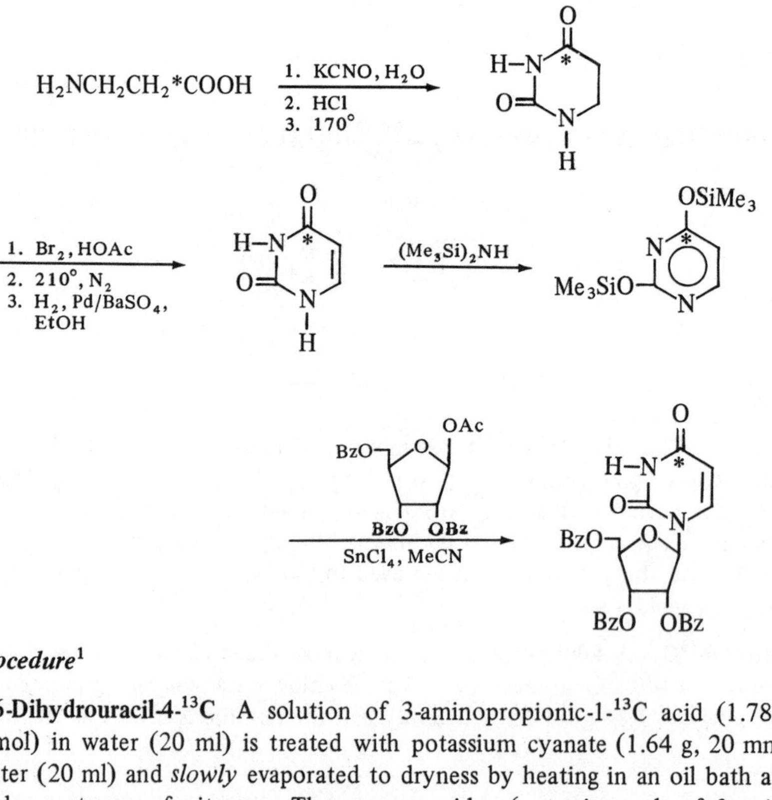

Procedure[1]

5,6-Dihydrouracil-4-^{13}C A solution of 3-aminopropionic-1-^{13}C acid (1.78 g, 20 mmol) in water (20 ml) is treated with potassium cyanate (1.64 g, 20 mmol) in water (20 ml) and *slowly* evaporated to dryness by heating in an oil bath at 100° under a stream of nitrogen. The syrupy residue (potassium salt of β-ureidopropionic-1-^{13}C acid), which solidifies on standing, is acidified with hydrochloric acid (6 *N*, 40 ml), evaporated to dryness, and heated at 170° for 30 min. The solid residue is extracted with water (5-ml portions) until all of the potassium chloride is removed, and the residue is dried over phosphorus pentoxide to afford the product (1.85 g). The water washings are extracted with ether for 72 hr to give an additional amount of crude product (total yield 2.27 g, 100% yield).

2,4-Bis(trimethylsiloxy)pyrimidine-4-^{13}C A stirred solution of 5,6-dihydrouracil-4-^{13}C (1.14 g, 10 mmol) in acetic acid is heated in an oil bath at 105° and treated over 3-4 hr with a solution of bromine (2.40 g, 15 mmol) in acetic acid (10 ml) at a rate to just maintain the bromine color (Note 1). The solvent is removed at reduced pressure to afford a white solid (Note 2), which is heated in an oil bath (preheated) at 210° for 25 min under a nitrogen stream to give a mixture of uracil-4-^{13}C and 5-bromouracil-4-^{13}C. The mixture is dissolved in aqueous ethanol (50%, 100 ml) by gentle heating and hydrogenated over palladium-on-barium sulfate (5%, 1.0 g) for 5 hr (Note 3). The solution is heated to boiling, the catalyst is filtered and washed with hot water (4 × 15 ml), and the filtrate is evaporated at reduced pressure to afford crude uracil-4-^{13}C (1.10 g, 99% yield). A suspension of the pale-yellow solid in hexamethyldisilazane (8 ml) is refluxed under nitrogen for 16 hr and concentrated at reduced pressure (90° at 50 mm Hg, then at 7 mm Hg), and the residue is distilled to afford the product as a colorless liquid, bp 125° at 5 mm Hg (2.40 g, 94% yield) (Note 4).

2',3',5'-Tri-*O*-benzoyluridine-4-^{13}C To a solution at 10° of the above-mentioned siloxy compound (2 g, 8.6 mmol) and 1-*O*-acetyl-2,3,5-tri-*O*-benzoyl-α-D-ribofuranoside in acetonitrile (100 ml) under nitrogen is added stannic chloride (1.6 g, 6.1 mmol) (Note 5) in acetonitrile (50 ml), stirred at 22° for 16 hr, and evaporated at reduced pressure. The residue is dissolved in ethylene chloride (300 ml), and the solution is shaken with sodium bicarbonate solution (saturated, 250 ml). The aqueous layer is repeatedly extracted with ethylene chloride (Note 6), and the extract is dried over sodium and magnesium sulfate and evaporated to afford a light, creamy white crystalline solid (4.81 g). Medium-pressure chromatography (silica gel, ICN 230-240 mesh, methylene chloride-methanol, 49:1) affords the product as a white crystalline solid, mp 142-143° (4.30 g, 90% yield).

Notes

1 The reaction flask is fitted with an efficient Dry Ice condenser.
2 ^{1}H-NMR confirms the absence of starting material and shows the solid to contain 5-bromodihydrouracil and 5,5-dibromohydrouracil in approximately equal amounts.
3 Theoretical uptake of hydrogen is complete after 2 hr.
4 The product is extremely moisture sensitive and is collected and stored under dry nitrogen.
5 The stannic chloride is freshly distilled from phosphorus pentoxide.
6 After several extractions the emulsion is broken by filtration through Whatman No. 1 paper.

Heterocyclic Compounds

References

[1]J. L. Roberts and C. D. Poulter, "2',3',5'-Tri-O-benzoyl[4-^{13}C]uridine. An Efficient, Regio-specific Synthesis of the Pyrimidine Ring," *J. Org. Chem.*, **43**, 1547-1550 (1978).

5-FLUORO-2'-DEOXYURIDINE-4-^{13}C

Procedure[1]

2,2'-Anhydro-β-D-arabinofuranosyluracil-4-^{13}C   A solution of methyl propiolate-1-^{13}C (6.7 g, 80 mmol) in tetrahydrofuran (15 ml) is added to 2-amino-β-D-arabinofurano[1',2':4,5]-2-oxazoline (13.9 g, 80 mmol) in water (70 ml) and ethanol (80 ml), and the mixture is refluxed for 5 hr. The volume is reduced, and the solution is cooled to give colorless crystals, which are filtered and washed with cold ethanol; mp 246-248° (10.7 g). Concentration of the mother liquor affords additional product (total yield 13.9 g, 77%).

3',5'-Di-O-acetyl-2'-bromo-2'-deoxyuridine-4-^{13}C The anhydro compound above (5.65 g, 24 mmol) is refluxed with acetyl bromide (9.23 g, 75 mmol) in dimethyl formamide (18 ml) and ethyl acetate (120 ml) for 1.5 hr. The cooled mixture is washed with water (3 × 50 ml), dried over sodium sulfate, and evaporated at reduced pressure to leave a pale-brown syrup, which is coevaporated with three small portions of ethyl acetate to give an amorphous solid, mp 55-57° (9.10 g, 97% yield).

3',5'-Di-O-acetyl-2'-deoxyuridine-4-^{13}C To a solution of the bromo compound (4.5 g, 11.5 mmol) in benzene (50 ml) is added tributyltin hydride (10.0 g, 35 mmol) in benzene (100 ml) and azobisisobutyronitrile (65 mg). The mixture is refluxed 1 hr, evaporated to 50 ml at reduced pressure, treated with hexane, and cooled. The viscous oil is triturated repeatedly with hexane until the odor of tributyltin compounds is no longer detectable, whereupon the oil crystallizes. Recrystallization affords colorless crystals, mp 108-109° (2.72 g, 76% yield).

5-Fluoro-2'-deoxyuridine-4-^{13}C The above acetylated uridine (0.43 g, 1.39 mmol) in chloroform (15 ml) is fluorinated at −78° with trifluoromethyl hypofluorite (0.30 g, 2.9 mmol) (Note 1) in trichlorofluoromethane (10 ml). After 1 hr at this temperature the excess hypofluorite is removed in a stream of nitrogen. The flask is covered with aluminum foil, and the solvents are removed at reduced pressure at 9°. The flocculent white solid is treated for 16 hr at 0° with a solution (20 ml) of triethylamine (10%) in aqueous methanol (50%). The mixture is applied to a column of Dowex-50(H⁺) (0.5 × 10 cm) and eluted with methanol. Fractions with significant absorption at 268 nm are combined and evaporated, and the residue is recrystallized from ethanol to afford the product, mp 148-150° (0.25 g, 73% yield) (Note 2).

Notes

1 Caution. The apparatus and technique used are described by Dawson and Dunlap.
2 Phosphorylation of the product with phosphorus oxychloride (no experimental details are given) affords 5-fluoro-2'-deoxyuridine-4-^{13}C 5'-monophosphate.

References

[1] W. H. Dawson and R. B. Dunlap, "An Improved Synthesis of 5-Fluoro-2'-deoxyuridine Incorporating Isotopic Labels," *J. Labelled Compds.*, **16**, 335-343 (1979).

THYMIDINE-6-^{13}C-α,α,α-d_3-^{15}N$_2$

$$CD_3CH(*CN)COOH + H_2*NCO*NH_2 \xrightarrow{Ac_2O}$$

$$CD_3CH(^*CN)CO^*NHCO^*NH_2 \xrightarrow[\text{HOAc, H}_2\text{O}]{D_2,\ PtO_2}$$

1. Me$_3$SiCl, Et$_3$N, C$_6$H$_6$

2. MeCN,
   p-MeC$_6$H$_4$COO

3. NaOMe, MeOH

Procedure[1]

1-[2-(Cyano-^{13}C)propionyl-3,3,3-d_3]urea-^{15}N$_2$ A mixture of 2-(cyano-^{13}C)propionic-3,3,3-d_3 acid (3.70 g, 36 mmol), urea-^{15}N$_2$ (2.50 g, 40 mmol), and acetic anhydride (4.5 ml, 45 mmol) is heated on a steam bath for 2 hr, treated with water (25 ml), heated for another 30 min, cooled, and filtered. The crystalline precipitate is washed with water (2 × 10 ml) and dried under vacuum to give the product, mp 190-191° (3.05 g, 57% yield, Note 1) (Note 2).

Thymine-6-^{13}C-α,α,α-d_3-^{15}N$_2$ A mixture of the above-mentioned ureide (2.41 g, 17 mmol), acetic acid (1.0 ml, 17 mmol), platinum oxide (0.20 g), and water (120 ml) is stirred under deuterium (1 atm) at 70° for 24 hr (Note 3), filtered, and concentrated at reduced pressure to a volume of 30 ml. The crystalline precipitate (0.87 g) is collected and dried. Further concentration to 15 ml gives a second crop (0.18 g) (total yield 1.05 g, 49%) (Note 4).

3′,5′-Di-O-p-toluoylthymidine-6-^{13}C-α,α,α-d_3-^{15}N$_2$ A mixture of the above-mentioned labeled thymine (0.91 g, 6.8 mmol), chlorotrimethylsilane (2.57 g, 23.6 mmol), triethylamine (2.6 ml, 18.8 mmol), and benzene (20 ml) is stirred at ambient temperature for 17 hr, filtered, and evaporated at reduced pressure. The residual bis-silylated thymine is dissolved in acetonitrile (15 ml) and treated with 1-chloro-3,5-di-O-p-toluoyl-2-deoxyribofuranose (2.65 g, 6.8 mmol). The mixture is stirred for 21 hr and evaporated at reduced pressure. The residue is partitioned between water and methylene chloride, and the methylene chloride extracts are washed with water, dried over magnesium sulfate, and evaporated. The foamy residue is recrystallized twice from methanol (12 ml) to afford the product, mp 197-198° (1.28 g, 39% yield) (Note 5).

Thymidine-6-^{13}C-α,α,α-d_3-^{15}N$_2$ A solution of the blocked nucleoside above (1.28 g, 2.6 mmol) in methanol (5 ml) containing sodium methoxide (30 mg) is refluxed under nitrogen for 19 hr. The solvent is evaporated at reduced pressure, and the residue is partitioned between water and ether. Evaporation of the aqueous solution affords a colorless gum, which is twice recrystallized from ethanol to give the product, mp 183-184° (0.42 g, 65% yield).

Notes

1 The yield of 57% is based on the acid; it is 52% based on the urea. Isotopic yield, based on total carbon-13 and nitrogen-15, is 54% [3·20.6/(36 + 2·40)].

2 NMR and field ionization MS data are given for products.

3 A total of 1.38 equiv. of gas is absorbed.

4 Thymine-6-^{13}C-α,α,α-d_3 is obtained in the same manner when ordinary urea is used as starting material. When reduction with deuterium of the ureide is carried out in water-d_2, thymine-6-^{13}C-$\alpha,\alpha,\alpha,6$-d_4 is obtained.

5 A second crop (0.5 g) appears to be a mixture of anomers inseparable by conventional chromatography.

References

[1]J. A. Lawson, J. I. DeGraw, and M. Anbar, "Synthesis of Hexalabeled Thymine and Thymidine," *J. Labelled Compds.*, 11, 489-499 (1975).

2-METHYLQUINOLINE-2-^{13}C

Procedure[1]

2-Methylquinoline-2-^{13}C To a stirred solution of sodium ethoxide (from sodium, 0.9 g, 39 mmol) in ethanol (100 ml) is added dropwise a solution of acetone-2-^{13}C (1.17 g, 19.8 mmol) and 2-aminobenzaldehyde (2.44 g, 20.2 mmol) in ethanol (50 ml). The mixture is refluxed for 12 hr, concentrated to 50 ml, added cautiously to cold water, adjusted to pH 7, and extracted with methylene chloride (5 × 50 ml). The extract is washed with water, dried over magnesium sulfate, and evaporated. The residue, by bulb-to-bulb distillation, bp 130-140° at 18 mm Hg, gives the product as a colorless liquid (2.79 g, 98% yield).

2-(Tribromomethyl)quinoline-2-^{13}C To a stirred solution of 2-methylquinoline-2-^{13}C (2.79 g, 19.4 mmol), sodium acetate (10 g), and acetic acid (20 g) is added bromine (9.6 g, 0.12 mol) over a period of 15 min. The mixture is refluxed for 1 hr, cooled, and poured into ice and water. The filtered precipitate is combined with additional material from methylene chloride (5 × 50 ml) extraction of the filtrate to afford a yellow solid, mp 129-130° (6.74 g, 91% yield).

References

[1]T. A. Bryson, J. C. Wisowaty, R. B. Dunlap, R. R. Fisher, and P. D. Ellis, "Biological Probes. I. Carbon-6-Labeled Nicotinammide," *J. Org. Chem.*, 39, 1158-1160 (1974).

$^{13}C_4$-CINOXACIN

Procedure[1]

6,7-(Methylenedioxy)-4-cinnolinol-3,4-$^{13}C_2$ A solution of sodium nitrite (1.35 g, 20 mmol) in water (5 ml) is added to a solution of 2′-amino-4′,5′-(methylene-dioxy)acetophenone-1,2-$^{13}C_2$ (2.7 g, 17.5 mmol) in hydrochloric acid (conc., 33 ml) in an ice bath, and the mixture is heated at 80° for 4 hr. The product is obtained as an ocherous powder (2.5 g, 87% yield).

3-Bromo-6,7-(methylenedioxy)-4-cinnolinol-3,4-$^{13}C_2$ A mixture of the product above (2.5 g, 15.2 mmol) and potassium acetate (1.27 g, 13 mmol) in acetic acid (17 ml) is treated with a solution of bromine (2 g, 12.5 mmol) in acetic acid (5 ml) to give a white powder (2.8 g, 80% yield).

4-Hydroxy-6,7-(methylenedioxy)-3-cinnolinecarbonitrile-*carbo*,3,4-$^{13}C_3$ A   mix-ture of the product above (1.7 g, 7.0 mmol) and cuprous cyanide-^{13}C (850 mg, 9.3

mmol) in dimethylformamide (20 ml) is refluxed for 4 hr to afford a yellow-green powder (1.2 g, 88% yield).

1-(Ethyl-1-^{13}C)-6,7-(methylenedioxy)-4(1H)-oxo-3-cinnolinecarbonitrile-*carbo*,3,4-^{13}C$_3$ Ethyl-1-^{13}C iodide (1.0 g, 6.4 mmol) is added to a brown solution obtained from a suspension of the above-mentioned nitrile (1.15 g, 6.0 mmol) and sodium hydride (154 mg, 6.4 mmol) in dimethylformamide (10 ml), and the solution is heated at 95-100° for 2 hr to give a yellow-brown powder (1.1 g, 84% yield).

1-(Ethyl-1-^{13}C)-6,7-(methylenedioxy)-4(1H)-oxo-3-cinnolinecarboxylic-*carboxy*,3,-4-^{13}C$_3$ acid (^{13}C$_4$-cinoxacin) The preceding nitrile (1.05 g, 4.75 mmol) is refluxed in a solution of hydrochloric acid (conc., 15 ml) and acetic acid (15 ml) for 2.5 hr to afford the product as pale-yellow crystals (from chloroform-methanol), mp 261-263° dec. (0.43 g, 38% yield).

Other Preparations

By the same procedure, starting from sodium nitrite-^{15}N and 2'-amino-4',5'-(methylenedioxy)acetophenone, is prepared 1-ethyl-6,7-(methylenedioxy)-4(1H)-oxo-3-cinnolinecarboxylic-2-^{15}N acid (^{15}N$_1$-Cinoxacin).[1]

References

[1]T. Nagasaki, Y. Katsuyama, and H. Minato, "Synthesis of Radioactive and Stable Isotope-Labelled 1-Ethyl-6,7-methylenedioxy-4(1H)-oxocinnoline-3-carboxylic Acids (Cinoxacin)," *J. Labelled Compds.*, **12**, 409-427 (1976).

THEOPHYLLINE-8-^{13}C

Procedure[1]

5,6-Diamino-1,3-dimethyluracil (2.0 g, 11.8 mmol) and formic-^{13}C acid (0.5 g, 10.6 mmol) are dissolved in water (15 ml), the pale-yellow solution is allowed to stand overnight, and the walls of the flask are scraped, which results in the formation of white crystals. After 3 days, during which time additional crystals form, a solution of sodium hydroxide (1 g, 25 mmol) in water (5 ml) is added, and the mixture is heated at 90° for 15 min, cooled in an ice bath, neutralized with hydrochloric acid, and kept at 4° overnight. The crystals are filtered and dissolved in hot water (30 ml), and the solution is treated with decolorizing carbon, filtered, and cooled to give the product (250 mg). Additional product in the filtrate is isolated by preparative thin-layer chromatography (silica gel; 20 cm × 20 cm × 2 mm; ether-methanol, 9:1, 7:3, and 2:8 in successive separations) and sublimed at less than 0.1 mm Hg to give pale-yellow material, which is treated with carbon and combined with the first crop. The combined material is purified by preparative thin-layer chromatography (silica gel, water) to give the product (514 mg, 27% yield, Note 1).

Notes

1 The yield could be increased by not using carbon and should still afford a product of suitable purity for many applications.

References

[1]C. E. Hignite, D. H. Huffman, D. L. Azarnoff, "Synthesis of Theophylline-8-^{13}C," *J. Labelled Compds.*, **14**, 475-478 (1978).

6,7-DIMETHYLLUMAZINE-8a-[13]C

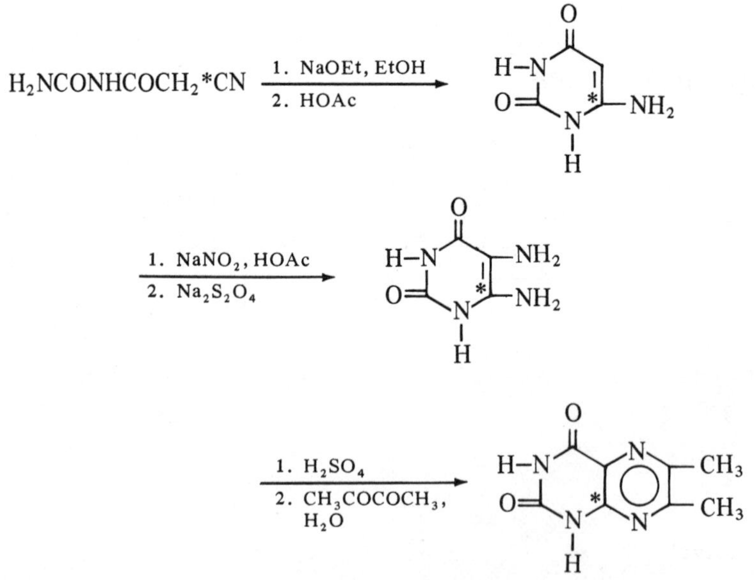

Procedure[1]

6-Aminouracil-6-^{13}C [(Cyano-^{13}C)acetyl]urea (2.9 g, 22.8 mmol) is refluxed in ethanol (30 ml) containing sodium ethoxide (from sodium, 2 g, 87 mmol) for 1 hr. Water (30 ml) is added, and the mixture is stirred until solution is effected, heated at 80° for 15 min, acidified with acetic acid to pH 5, and allowed to cool to afford a white solid (2.68 g, 92% yield).

5,6-Diaminouracil-6-^{13}C A suspension of the product above (2.68 g, 21.1 mmol) in water (20 ml) is acidified with acetic acid, treated with sodium nitrite (1.78 g, 25.8 mmol), and heated at 80° for 15 min. To the resulting, stirred suspension of rose-colored 6-amino-5-nitrosouracil-6-^{13}C at 80° is added sodium hydrosulfite until the color is completely bleached, and heating is continued for an additional 15 min. The cooled mixture affords the product as a buff-colored precipitate (2.80 g, 93% yield). The hemisulfate is obtained by pouring a solution of the crude product in dilute sodium hydroxide (5 ml) into boiling dilute sulfuric acid. The crystalline hemisulfate precipitates on cooling of the solution.

6,7-Dimethyllumazine-8*a*-^{13}C The hemisulfate above (5.0 g, 26.2 mmol) is heated at 100° in water (30 ml) with 2,3-butanedione (2.25 g, 26.1 mmol) for 15 min to afford the product as a crystalline solid (3.34 g, 67% yield).

References

[1]J. W. Triplett, S. W. Mack, S. L. Smith, and G. A. Digenis, "Synthesis of Carbon-13 Labelled Uracil, 6,7-Dimethyllumazine and Lumichrome, via a Common Intermediate: Cyanoacetylurea," *J. Labelled Compds.*, **14**, 35-41 (1978).

LUMICHROME-10*a*-^{13}C

Procedure[1]

A mixture of 5,6-diaminouracil-6-^{13}C hemisulfate (0.80 g, 4.16 mmol) and 5-acetyl-tetrahydro-2-hydroxy-2,5-dimethyl-5-oxofuran (0.716 g, 4.16 mmol) in water (15 ml) is heated at $100°$ for 1 hr. The suspension is filtered and cooled to afford pale-yellow crystals of 2,4-dihydroxy-6-(2-hydroxy-2-methyl-3-oxobutyl)-7-methylpteri-dine-8*a*-^{13}C, which are dissolved in sodium hydroxide (2 *N*, 20 ml) and heated on a steam bath for 1 hr. The solution is cooled to give the product as the bright-yellow sodium salt (485 mg, 44% yield).

References

[1]J. W. Triplett, S. W. Mack, S. L. Smith, and G. A. Digenis, "Synthesis of Carbon-13 Labelled Uracil, 6,7-Dimethyllumazine and Lumichrome, via a Common Intermediate: Cyanoacetyl-urea," *J. Labelled Compds.*, **14**, 35-41 (1978).

9*H*-XANTHEN-9-ONE-9-[18]O

Procedure[1]

A solution of 9*H*-xanthene (0.1 g, 0.55 mmol) in 3-methylpentane (25 ml) con-tained in a 50-ml quartz bulb is degassed by passing nitrogen through the solution for 15 min followed by the freeze-pump-thaw method. The system is evacuated, the flask is cooled in liquid nitrogen, and oxygen-[18]O_2 (approx. 50 ml, approx. 2 mmol) is introduced. The solution is saturated with the gas at approx. 380 mm Hg by vigorous magnetic stirring for 15 min. The reaction flask is closed by a stopcock and irradiated for 1 hr with a focused, heat-filtered xenon lamp (Oriel, 1500 watt) at a distance of 50 cm. The solution is evaporated at reduced pressure, and the resi-due in benzene-pentane (1:1, 2 ml) is applied to a chromatography column (silicic acid, 1 X 30 cm) (Note 1) to afford the product (10 mg, 0.05 mmol) (Note 2).

Notes

1 The effluent is monitored by measuring the absorbance at 338 nm for product and at 292 nm for starting material.
2 All operations are performed in subdued light and in Pyrex containers to avoid photooxidation. Relative intensities of the parent peaks at masses 196 and 198 indicate the isotopic purity of the product is 95% before and 80% after chroma-tography. If higher isotopic purity is desired, purification should be carried out under a nitrogen atmosphere.

References

[1]H. J. Pownall, "Photo-Oxidation in the Synthesis of ^{17}O and ^{18}O Labelled Compounds: Synthesis of Xanthone-^{18}O$_1$," *J. Labelled Compds.*, **10**, 413-417 (1974).

7

Other Compounds

Many of the remarks in the introductory paragraphs to the preceding chapters also apply to compounds that are included here. Nuclear magnetic resonance spectroscopy continues to be increasingly valuable to the synthetic chemist in developing new or improved methods. Application has been made[1,2] to improving and better understanding the procedures for labeling sugars by the Kiliani-Fischer type of reaction. The possibilities of applying the Sowden and related syntheses, using nitromethane, have been discussed;[3] it is likely they will be used more extensively in the future to introduce isotopes into carbohydrates—a most important class of labeled compounds.

References

[1] R. M. Blazer and T. W. Whaley, "A Carbon-13 Nuclear Magnetic Resonance Spectroscopic Investigation of the Kiliani Reaction," *J. Am. Chem. Soc.*, in press (1980).

[2] A. S. Serianni, H. A. Nunez, and R. Barker, "Carbon-13-Enriched Carbohydrates. Preparation of Aldononitriles and Their Reduction with a Palladium Catalyst," *Carbohydr. Res.*, 72, 71-78 (1979).

[3] L. Benzing-Nguyen and M. B. Perry, "Stepwise Synthesis of N-Acetylneuraminic Acid and N-Acetyl[1-^{13}C]neuraminic Acid," *J. Org. Chem.*, 43, 551-554 (1978).

CARBON-^{13}C

$$\text{*CO}_2 \xrightarrow[\text{550-600}^\circ]{\text{H}_2, \text{Fe}} \text{*C}$$

Procedure[1]

The apparatus, which provides for continuous circulation of the reactant gases at 1-2 liter/min over the powdered iron catalyst (60 mg) at a temperature 550-600° and a pressure of 800 mm Hg (Note 1), is filled initially with carbon-^{13}C dioxide (320 mm Hg) and hydrogen (480 mm Hg). After about 1 hr a steady rate is attained corresponding to consumption of approx. 1 liter of carbon dioxide/hr, and hydrogen and carbon dioxide are added as the reaction proceeds. When the carbon dioxide supply is exhausted (1.20 STPl, 53.6 mmol) (Note 2), the iron catalyst is removed by reaction with hydrogen chloride and increase of the temperature to 900°. A mixture (10 liter) of argon and hydrogen chloride is passed through the reaction tube at approx 180 ml/min over a period of 1-2 hr (Note 3). The yield of purified product is 0.704 g (98%).

Notes

1 A diagram of the apparatus is given by Rutherford and Liner. The optimum concentration of hydrogen is 40-60% and is maintained with the aid of a heated palladium-silver alloy diffusion thimble.

2 Excess hydrogen pressure remains. A small portion of the carbon is present in the gas phase as methane.

3 Iron is volatilized as the chloride, which is transported downstream to colder parts of the tube. The process is repeated several times to remove the iron completely.

References

[1]W. M. Rutherford and J. C. Liner, "Preparation of Elemental ^{13}C and K^{13}CN from ^{13}CO$_2$," *Int. J. Appl. Radiat. and Isot.*, **21**, 71-73 (1970).

CARBON-^{13}C OXYSULFIDE

$$*CO \xrightarrow[\text{MeOH, 100°}]{\text{S, NaOMe}} *COS$$

Procedure[1]

Sulfur (32 g, 1.0 mol), sodium methoxide (2 g, 37 mmol), and methanol (50 ml) are placed in a 1-liter stainless steel cylinder equipped with a thermometer, pressure gauge, and heating jacket. The vessel is partially evacuated, carbon monoxide (200 psi, 12.9 g, 0.45 mol) is added, and the mixture is shaken and heated to 100-105° (Note 1). After 20 hr the vessel is cooled to room temperature and vented through two traps at −40°, then a tube packed with Lithasorb, and finally through traps at Dry Ice temperature (Note 2). The product is cryogenically transferred from the Dry Ice traps to a 300-ml stainless steel storage cylinder (19.6 g, 71% yield).

Notes

1 The pressure increases to 250-260 psi and then falls and levels off at near 200 psi.

2 The traps at −40° remove entrained methanol, and the Lithasorb absorbs a small amount of by-product carbon dioxide. Most of the product, bp −50°, is contained in the first Dry Ice trap.

References

[1]V. N. Kerr and D. G. Ott, "Preparation of Carbon-^{13}C Disulfide and Carbon-^{13}C Oxysulfide," *J. Labelled Compds.*, **14**, 793-795 (1978).

CARBON-^{13}C DISULFIDE

$$*CH_4 \quad \xrightarrow[\substack{Quartz, \\ 1000°}]{H_2S} \quad *CS_2$$

Procedure[1]

A quartz combustion tube (2.4 × 90 cm) is packed along 70 cm of its length with quartz tubing (5 mm o.d., 3 mm i.d. × 25 cm) (17 pieces in cross section, Note 1) and heated in an electric furnace to 975-1000°. A mixture of methane-^{13}C (33.8 g, 2.00 mol) (at approx. 100 ml/min) and excess hydrogen sulfide (at approx. 275 ml/min) is passed through the hot tube over about 9 hr, and the exit gas is passed through an air-cooled condenser and then through a Dry Ice condenser fitted to a flask cooled to −20° (Note 2). The flask contents are redistilled to afford the product, bp 39.5° at 590 mm hg (53.4 g, 72% yield).

Notes

1   The pieces of quartz tubing afford approx. 3500 cm^2 of quartz surface area; the free volume is approx. 250 cm^3.

2   Dry Ice traps are attached to the outlet from the flask; however, essentially all the product is retained in the flask.

References

[1]V. N. Kerr and D. G. Ott, "Preparation of Carbon-^{13}C Disulfide and Carbon-^{13}C Oxysulfide," *J. Labelled Compds.*, **14**, 793-795 (1978).

POTASSIUM NITRITE-$^{18}O_2$

$$KNO_2 \quad \xrightarrow[Dowex-50]{H_2*O} \quad KN*O_2$$

Procedure[1] (Note 1)

Potassium nitrite (2.5 g, 29 mmol) in stirred ice-cold water-^{18}O (10 ml, 0.51 mol) is treated with dry Dowex-50W(H$^+$) (0.5 g), and the mixture is allowed to warm to room temperature, kept for 24 hr, and centrifuged. The supernatant solution is brought to pH 4.9 with potassium hydroxide in water-^{18}O. The water is removed at reduced pressure to afford the product (Note 2).

Notes

1 Customary exchange reactions catalyzed by mineral acids are not applicable to nitrite because of the instability of nitrous acid. Use of a cation exchange resin (H$^+$ form) to catalyze the reaction obviates various difficulties. The procedure can be used for other salts (cations). No exchange has been observed using the anion exchange resin Dowex-1.

2 The product contained 73.94 at. % ^{18}O, and the recovered water contained 75.12 at. % ^{18}O, which indicates almost complete isotopic exchange.

Other Preparations

Sodium nitrite-$^{18}O_1$ has been prepared from dinitrogen trioxide-$^{18}O_1$ and sodium hydroxide. The oxide is obtained by oxidation of nitric oxide with oxygen-$^{18}O_2$. Through an analogous series of reactions have been prepared dinitrogen-$^{15}N_2$ trioxide (from nitric-^{15}N oxide, obtained from nitric-^{15}N acid and ferrous sulfate), sodium nitrite-^{15}N, and nitrogen-^{15}N dioxide. Nitrogen dioxide-$^{18}O_1$, from nitric oxide and oxygen-$^{18}O_2$, is oxidized further with oxygen in the presence of water to nitric-$^{18}O_1$ acid. These intermediates have been used in the preparation of nitromethane-$^{18}O_1$, nitromethane-^{15}N, methyl nitrate-N-$^{18}O_1$, methyl-$^{18}O_1$ nitrate, and methyl nitrate-^{15}N.[2]

References

[1]D. Samuel and I. Wassermann, "A New Method for the Synthesis of ^{18}O-Labelled Potassium Nitrite and Related Compounds," J. Labelled Compds., 7, 355-356 (1971).

[2]A. P. Cox and S. Waring, "Preparation of Methyl Nitrate and Nitromethane Labelled with Nitrogen-15 and Oxygen-18," J. Labelled Compds., 9, 153-157 (1973).

NITRIC-$^{18}O_3$ ACID

$$HNO_3 \xrightarrow{H_2*O} HN*O_3 \xrightarrow{NH_3} NH_4N*O_3 \xrightarrow{MeSO_3H} HN*O_3$$

Procedure[1]

Nitric-$^{18}O_3$ acid (39 at. % ^{18}O, 7.5 g, 0.12 mol) (Note 1) is allowed to exchange with water-^{18}O (95 at. % ^{18}O, 13 ml, 0.65 mol) at 75° (Note 2). After 6 days (Note 3) the solution is neutralized with ammonia, the water is removed by vacuum distillation, the residual ammonium nitrate-$^{18}O_3$ is dissolved in methanesulfonic acid (Note 4), and the nitric-$^{18}O_3$ acid is distilled into a liquid nitrogen-cooled trap under vacuum (75 at. % ^{18}O, 5.7 g, 88 mmol). A second exchange with water-^{18}O (98 at. % ^{18}O, 13 ml) for 14 days (Note 5) and treatment as above affords the anhydrous product (89 at. % ^{18}O, 4.3 g, 66 mmol) (Note 6).

Notes

1   The starting material in this example is isotopically labeled owing to its having been produced in another experiment. When other isotopic concentrations are used, the proportions of reactants are chosen so as to maximize oxygen transfer to nitric acid while allowing a reasonable rate.

2   The temperature is chosen to minimize thermal decomposition of nitric acid.

3   Progress of the exchange is followed by evaporation of aliquots (0.05 ml) that have been neutralized with ammonia, conversion of the salt to nitric acid by methanesulfonic acid, addition of anisole in methylene chloride, and analysis of the 4-(nitro-^{18}O$_3$)anisole by field ionization mass spectrometry (parent ion mass peaks at 153, 155, 157). Oxygen isotopic abundance in the water is determined by reaction with methyl sulfate and analysis of the resulting methanol-^{18}O (masses 32 and 34).

4   The rate of oxygen exchange between nitric acid and methanesulfonic acid is quite slow. Exchange reactions with oleum, sulfuric acid, and trifluoroacetic acid are rapid.

5   A longer exchange period is used because of the lower nitric acid concentration.

6   Complete isotopic exchange would give 91.8 at. % ^{18}O.

References

[1]A. C. Scott, J. H. McReynolds, and M. Anbar, "The Synthesis of ^{18}O Multilabeled Anhydrous Nitric Acid," *J. Labelled Compds.*, **12**, 63-67 (1976).

^{13}C-METHYLTIN IODIDES

$$*CH_3I \xrightarrow[\substack{2.\ CO_2 \\ 3.\ HI}]{1.\ K_2SnO_2,\ H_2O,\ EtOH} *CH_3SnI_3$$

$$*CH_3SnI_3 + 2Me_4Sn \xrightarrow{180°} *CH_3Me_2SnI + 2Me_3SnI$$

Procedure[1]

Methyltin-^{13}C triiodide   A solution of potassium hydroxide (3.3 g, 58.9 mmol) in water (13 ml) is cooled to 0° and slowly added to a stirred solution of stannous chloride (1.40 g, 7.38 mmol) in water (3.0 ml) at 0°. To the filtered potassium stannite solution is added methyl-^{13}C iodide (1.00 g, 7.04 mmol) in ethanol (3.0 ml) at 0° with the aid of more alcohol (3 × 1.0 ml). The mixture is magnetically stirred for 2 hr at 0° (Note 1) and then approx 13 hr at room temperature. Carbon dioxide is passed through the solution maintained at 0° by addition of Dry Ice for 2 hr (Note 2), and volatile components are removed at reduced pressure (approx. 20 mm Hg and 50°). Water (10 ml) and carbon tetrachloride (10 ml) are added to the white solid residue, hydriodic acid (conc.) is added to the stirred mixture until no solid remains, and the layers are separated. The aqueous layer is extracted with carbon tetrachloride (5 X 10 ml), and the combined extracts are washed with water, dried over sodium sulfate, and evaporated at reduced pressure. The yellow solid is recrystallized from light petroleum (bp 40-60°) to give the pure product. Repeated concentration and extraction of the aqueous layers afford additional recrystallized product (total yield 2.54 g, 70%) (Note 3).

Trimethyltin-^{13}C$_1$ iodide Methyltin-^{13}C triiodide (1.00 g, 1.94 mmol) and tetramethyltin (0.70 g, 3.91 mmol) are heated in a 5-ml flask under nitrogen in a bath at 180° until refluxing ceases (approx. 4.5 hr) and then for a further 3 hr. Distillation gives a small amount of tetramethyltin, bp approx. 75°, and then the pure product as a very pale-yellow liquid, bp 160-180° (1.42 g, 85% yield) (Note 4).

Notes

1 After about 1 hr the methyl iodide has dissolved.
2 A white finely crystalline precipitate is produced.
3 Evaporation of the final mother liquor gives additional crude product (0.62 g, 87% total yield).
4 Analysis by ^{1}H-NMR shows the expected isotopic concentration (approx. oneninth that of the starting material).

References

[1]J. D. Kennedy, "The Small-Scale Synthesis of ^{13}C-Labelled Methyltin Iodides," *J. Labelled Compounds*, 11, 285-286 (1975).

PHENYLISOTHIOCYANATE-^{15}N

$$Ph*NH_2 \xrightarrow[\text{2. EtOOCCl, Et}_3\text{N, CHCl}_3]{\text{1. CS}_2\text{, Et}_3\text{N, C}_6\text{H}_6} Ph*NCS$$

Procedure[1]

A solution of aniline-^{15}N (1.28 g, 13.6 mmol) in benzene (5 ml) is treated with carbon disulfide (1.1 ml, 18 mmol) and triethylamine (2.3 ml, 17 mmol). After 1 day the precipitated triethylammonium salt of *N*-phenyldithiocarbamic-^{15}N acid is filtered, washed with benzene, dried (3.53 g, 99% yield), dissolved in chloroform (15 ml), and treated at 0° with triethylamine (2 ml, 14.5 mmol) and ethyl chloroformate (1.5 ml, 14 mmol). After 1 hr the mixture is extracted with hydrochloric aicd (3 *M*, 2 X 10 ml), dried, and evaporated. The residue is decomposed at 130-

140° and fractionated to afford the product (containing a trace of ethanol) as a colorless oil (1.8 g, 97% yield).

References

[1]J. Volford and D. Banfi, "Synthesis of ^{15}N-Labeled 2-Substituted 2-Thiazolines and Analogous Thiazines," *J. Labelled Compds.*, **11**, 419-426 (1975).

4-TOLUENESULFONYLMETHYL(ISOCYANIDE-^{13}C) (^{13}C-TosMIC)

$$H*COONa \xrightarrow[190-250°]{CH_3NH_2 \cdot HCl} H*CONHCH_3$$

$$\xrightarrow[\substack{2.\ BuLi,\ THF,\ C_5H_{12},\ -70° \\ 3.\ 4-CH_3C_6H_4SO_2F,\ THF,\ -60°}]{\substack{1.\ 4-CH_3C_6H_4SO_2Cl, \\ quinoline}} 4-CH_3C_6H_4SO_2CH_2N*C$$

Procedure[1]

A mixture of sodium formate-^{13}C (1.3 g, 19 mmol) and methylamine hydrochloride (1.3 g, 20 mmol) is heated in a metal bath at 190°. The apparatus consists of a flask fitted with a cold-finger condenser having a small receiver attached at the bottom, which retains the condensate (Note 1). The bath temperature is slowly raised to 250°, and the *N*-methylformamide-^{13}C and water that form condense on the cold finger and deposit in the receiver (Note 2).

To a vigorously stirred solution at 75° of 4-toluenesulfonyl chloride (8 g, 42 mmol) and quinoline (14 g, 0.11 mol) evacuated to 15 mm Hg is added from a pressure-equalized funnel the amide above at a rate that maintains a smooth distillation rate (approx. 45-60 min). The methyl isocyanide-^{13}C, which collects in a liquid nitrogen-cooled receiver, is redistilled at atomspheric pressure, bp 59-60°.

To a solution of the isocyanide in tetrahydrofuran (25 ml) is added with vigorous stirring at −70° butyllithium in pentane (2 *M*, 10 ml). To the suspension at −70° is quickly added 4-methylbenzenesulfonyl fluoride (1.75 g, 10 mmol) in tetrahydrofuran (8 ml). The mixture is stirred at −60 to −50° for 10 min, treated with acetic acid (0.6 g, 10 mmol), and evaporated at 25° at reduced pressure. The residue is treated with water (13 ml) and extracted with methylene chloride (2 × 20 ml), and the extract is dried over magnesium sulfate and evaporated at reduced pressure to afford the product (1.72 g, 31% yield).

Notes

1 A diagram of the apparatus is given by Gossauer and Suhl.

2 To transfer completely the reaction product, *N*-methylformamide (0.5 g) is added to the residue and allowed to distill into the receiver.

References

[1] A. Gossauer and K. Suhl, "Total Synthesis of Verrucarin E. Its Application to Preparation of a ^{13}C-Labeled Derivative," *Helv. Chim. Acta,* 59, 1698-1704 (1976).

2-AMINOETHANESULFONIC-$^{13}C_1$ ACID

$$Na*CN + ClCH_2OCOCMe_3 \xrightarrow{\text{DMSO}} Me_3CCOOCH_2*CN$$

$$\xrightarrow[\substack{\text{2. HCl, H}_2\text{O} \\ \text{3. NH}_3\text{, EtOH}}]{\text{1. BH}_3\text{, THF}} HOCH_2*CH_2NH_2 \xrightarrow[\text{H}_2\text{O}]{\text{HBr}}$$

$$BrCH_2*CH_2NH_2 \cdot HBr \xrightarrow[\text{2. HCl}]{\text{1. Na}_2\text{SO}_3\text{, H}_2\text{O}} H_2N^xCH_2{}^xCH_2SO_3H$$

Procedure[1]

(Cyano-^{13}C)methyl pivalate   A stirred slurry of sodium cyanide-^{13}C (4.65 g, 93 mmol) in dimethyl sulfoxide (60 ml) is treated slowly with chloromethyl pivalate (12.5 g, 83 mmol). After one-third of the ester is added the temperature is raised to 55-60° and then maintained there by the rate of addition. Stirring is continued as the mixture returns to room temperature, water (250 ml) is added, and the solution is extracted with dichloromethane (4 × 75 ml). The extract is dried over sodium sulfate and evaporated to give an oil, which is distilled at reduced pressure, bp 90-91.5° at 21 mm Hg (9.4 g, 71% yield).

2-Aminoethanol-2-^{13}C   A solution of (cyano-^{13}C)methyl pivalate (4.0 g, 28 mmol) in tetrahydrofuran (20 ml) is added from a syringe over 30 min to a solution of borane in tetrahydrofuran (1 M, 100 ml, 100 mmol) under nitrogen. The mixture is refluxed 1 hr, cooled, treated slowly with hydrochloric acid (6 N, 50 ml), and evaporated. The residue is triturated with water, and the mixture is filtered and evaporated at reduced pressure. The residue is triturated with ethanol (20 ml) saturated with ammonia, and the solution is filtered and evaporated. The residual oil is dissolved in hydrochloric acid (1.5 N, 10 ml), and the solution is filtered and evaporated to a paste, which is triturated with ethanol saturated with ammonia. The solution is filtered, the filtrate is evaporated, the residue is dissolved in isopropyl alcohol (5 ml) and ethanol (5 ml), and the solution is saturated with hydrogen chloride. Ether (20 ml) is added to give an oil that crystallizes to afford the hydrochloride as a white solid, mp 78-80° (1.95 g, 73% yield). The free base is regenerated by treatment of the hydrochloride with ammonia-saturated ethanol, and the solution is filtered to remove ammonium chloride and evaporated.

2-Bromoethylamine-1-^{13}C hydrobromide   2-Aminoethanol-2-^{13}C (10.1 g, 0.174 mol) is treated with ice-cold hydrobromic acid (48%, 90 ml) and left at room temperature overnight. The mixture is refluxed 1 hr, distillate (22 ml) is removed, and the process of alternately refluxing for 1-hr periods and removal of distillate is repeated so as to remove successively 7-, 5-, 3-, 2-, and 1-ml portions. Hydrobromic acid (48%, 35 ml) is added and distilled, and the mixture is cooled to approx. 35°, treated with acetone (55 ml), and chilled for several hours to afford white crystals, mp 173-174° (12.38 g). Successive concentrations of the filtrate provide additional crops of product (total yield 19.7 g, 55%).

Taurine-^{13}C$_1$ 2-Bromoethylamine-1-^{13}C hydrobromide (18.9 g, 92 mmol) in water (45 ml) is added to sodium sulfite (12.8 g, 0.1 mol) in water (30 ml) in a 125-ml Erlenmeyer flask, which is heated on a steam bath for 12 hr. The mixture, now approximately one-third of the initial volume and containing considerable solid, is treated with hydrochloric acid (conc., 10 ml), cooled, and filtered. The solid is washed with hydrochloric acid (conc., approx. 6 × 30 ml) until no further material appears to dissolve. The filtrate is evaporarted to approx. 50 ml, heated to dissolve the solid, treated with hot ethanol (150 ml), and cooled. The solid is filtered, washed with ethanol, and dried (10.3 g, 89% yield, Note 1) (Note 2).

Notes

1   The overall yield from cyanide is 25%.
2   ^{13}C-NMR shows the product to be a mixture of 28% taurine-1-^{13}C and 72% taurine-2-^{13}C indicating that 56% of the reaction proceeds through the symmetrical aziridine-2-^{13}C intermediate.

References

[1]G. H. Daub, V. N. Kerr, D. L. Williams, and T. W. Whaley, "Organic Synthesis with Stable Isotopes," in *Proceedings of the Third International Conference on Stable Isotopes, Argonne National Laboratory*, E. R. Klein and P. D. Klein, Eds., Academic Press, New York, 1979, pp. 7-10.

2-AMINOETHANESULFONIC-^{13}C$_2$ ACID

$$\text{Br*CH}_2\text{*CH}_2\text{Br} \xrightarrow[\text{EtOH, H}_2\text{O}]{\text{Na}_2\text{SO}_3} \text{Br*CH}_2\text{*CH}_2\text{SO}_3\text{Na}$$

$$\xrightarrow{\text{NH}_4\text{OH}} \text{H}_2\text{N*CH}_2\text{*CH}_2\text{SO}_3\text{H}$$

Procedure[1]

Sodium 2-bromoethanesulfonate-^{13}C$_2$ To a stirred refluxing solution of 1,2-dibromoethane-^{13}C$_2$ (50.0 g, 0.265 mol) in ethanol (100 ml) and water (37 ml) is added a solution of sodium sulfite (11.0 g, 87 mmol) over 2 hr. After 4 hr the ethanol and dibromoethane are distilled, and the aqueous fraction is evaporated to give a residue that is extracted with hot ethanol (2 × 100 ml). Filtration and evaporation of the solvent afford the product. The recovered distillate of ethanol and dibromoethane is again reacted with sodium sulfite in water, and the procedure is repeated until the dibromethane is consumed. Five such treatments afford the product, mp 283-285° dec. (45.8 g, 81% yield).

Taurine-$^{13}C_2$ The above product (0.218 mol) is dissolved in ammonium hydroxide (conc., 500 ml), and the mixture is allowed to stand for one week to afford the product (21.8 g, 80% yield) (Note 1 and 2).

Notes

1. IR and TLC data are given.
2. Treatment of the product with benzyl chloroformate followed by the usual procedures affords N-benzyloxycarbonyltaurine-$^{13}C_2$ amide, which is amenable to MS analysis.

References

[1] B. D. Andresen, F. T. Davis, P. P. Toskes, and C. E. King, "Synthesis of Taurine-$^{13}C_2$. A Potential Probe for Human $^{13}CO_2$ Breath Tests," *J. Labelled Compds.*, **15**, 731-738 (1978).

4-METHYLPHENYL 1-NAPHTHYL SULFONE-$^{18}O_1$

Method I

$$4-\text{MeC}_6\text{H}_4-\underset{\underset{O}{\|}}{S}-\text{C}_{10}\text{H}_7-1 \quad \xrightarrow[\substack{\text{PhICl}_2, \\ \text{AgNO}_3}]{\substack{\text{H}_2\text{*O}, \\ \text{C}_5\text{H}_5\text{N}}} \quad 4-\text{MeC}_6\text{H}_4-\overset{\text{*O}}{\underset{\underset{O}{\|}}{\overset{\|}{S}}}-\text{C}_{10}\text{H}_7-1$$

Procedure[1]

(−)-4-Methylphenyl 1-naphthyl sulfone-$^{18}O_1$ A solution of (dichloroiodo)benzene (0.55 g, 2 mmol) in pyridine (10 ml) is added to a stirred solution of (−)-4-methylphenyl 1-naphthyl sulfoxide (0.54 g, 2 mmol), water-^{18}O (0.25 ml, 12.5 mmol), and silver nitrate (0.81 g, 4.8 mmol) in pyridine (5 ml) at −40°. The mixture is kept at −40° for 1 hr and then at room temperature for 1 hr and purified twice by column chromatography (silica; ether-light petroleum, 1:1) to give the product, mp 123-124° (0.42 g, 74% yield) (Note 1).

Method II

$$4-\text{MeC}_6\text{H}_4-\underset{\underset{O}{\|}}{S}-\text{C}_{10}\text{H}_7-1 \quad \xrightarrow[\substack{\text{2. Na*OD, D}_2\text{*O}, \\ \text{dioxane}}]{\substack{\text{1. Et}_3\text{O}^+\text{BF}_4^-, \\ \text{CH}_2\text{Cl}_2}} \quad 4-\text{MeC}_6\text{H}_4-\overset{\text{*O}}{\underset{\underset{O}{\|}}{\overset{\|}{S}}}-\text{C}_{10}\text{H}_7-1$$

$$\xrightarrow[\text{CH}_2\text{Cl}_2]{\text{3-ClC}_6\text{H}_4\text{CO}_3\text{H}} \quad 4-\text{MeC}_6\text{H}_4-\overset{\overset{*\text{O}}{\|}}{\underset{\|}{\underset{\text{O}}{\text{S}}}}-\text{C}_{10}\text{H}_7-1$$

Procedure[1]

(+)-4-methylphenyl 1-naphthyl sulfoxide-^{18}O (−)-4-Methylphenyl 1-naphthyl sulfoxide, $[\alpha]_D^{25}$ −399° (c 1, acetone), (0.67 g, 2.5 mmol) and triethyloxonium fluoroborate (0.475 g, 2.5 mmol) in methylene chloride (approx. 10 ml) are stirred at room temperature for 2 days. The resulting oil is washed several times with ether and benzene and hydrolyzed in dioxane (6 ml) with sodium deuteroxide (5 mmol) in water-d_2-^{18}O (0.6 ml) at room temperature for 24 hr. The mixture is diluted with chloroform (20 ml), dried over sodium sulfate, and evaporated at reduced pressure to give the product, mp 136° (from cyclohexane), $[\alpha]_D^{25}$ +380° (c 1, acetone) (0.47 g, 70% yield) (Note 2).

(+)-4-Methylphenyl 1-naphthyl sulfone-^{18}O$_2$ The sulfoxide above (0.27 g, 1 mmol) is oxidized with 3-chloroperoxybenzoic acid (slight excess) in dichloromethane at room temperature for 24 hr. The crude product is chromatographed twice (silica; ether-light petroleum, 1:1) to give the pure sulfone, mp 124° (from methanol) (0.22 g, 77% yield) (Note 3).

Notes

1 The oxidation proceeds through nucleophilic attack at tetracoordinate hexavalent sulfur and involves overall inversion of configuration. Without silver nitrate the reaction is very slow, and the product is racemized, probably because of chloride ions. The water is 92 at. % ^{18}O, and the product is 75 at. % ^{18}O.
2 The sulfoxide has an isotopic concentration of 80 at. % ^{18}O; the water-d_2-^{18}O is 92 at. % ^{18}O.
3 The sulfone is 83 at. % ^{18}O at the labeled position.

Other Preparations

By the procedure of Method II, (+)-benzyl 4-methylphenyl sulfoxide is converted to (−)-benzyl-α,α-d_2 4-methylphenyl sulfoxide-^{18}O (88% yield) and oxidized to (−)-benzyl-α,α-d_2 4-methylphenyl sulfone-^{18}O$_1$ (80% yield).[1]

References

[1]R. Annunziata, M. Cinquini, and S. Colonna, "Synthesis and Stereochemistry of Optically Active ^{16}O, ^{18}O Sulphones," *J. Chem. Soc.*, 1972, 2057-2059.

TRIMETHYLOXOSULFONIUM-^{13}C$_1$ IODIDE

$$*\text{CH}_3\text{I} + (\text{CH}_3)_2\text{SO} \longrightarrow *\text{CH}_3(\text{CH}_3)_2\text{SO}^+\text{I}^-$$

Procedure[1]

Dimethyl sulfoxide (8.25 g, 106 mmol) (Note 1) and methyl-^{13}C iodide (9.8 g, 69 mmol) (Note 2) are refluxed under nitrogen in a mercury-sealed system for 24 hr. The precipitated yellowish-brown solid is filtered, washed with cold tetrahydrofuran (20 ml), and air dried to give the product as fine white crystals, mp 176-177° (11.7 g, 77% yield).

Notes

1 The dimethyl sulfoxide is dried over calcium hydride and distilled at reduced pressure.
2 In trial experiments the ratio of reactants has been varied over a wide range; the molar ratio of 1:1.5 gives the maximum yield.

References

[1]M. A. G. El-Fayoumy, H. C. Dorn, and M. A. Ogliaruso, "Preparation of Trimethyloxosulfonium-^{13}C Iodide," J. Labelled Compds., 13, 433-436 (1977).

4-ETHYLSULFONYL-1-NAPHTHALENESULFONAMIDE-^{15}N (^{15}N-ENS)

Procedure[1] (Note 1)

4-Ethylthio-1-naphthalenesulfonamide-^{15}N Ammonium-^{15}N sulfate (1.21 g, 9.0 mmol), 4-ethylthio-1-naphthalenesulfonyl chloride (5.16 g, 18 mmol), potassium carbonate (9.94 g, 72 mmol), and acetonitrile (72 ml) are cooled in an ice bath, water (36 ml) is added, the flask is stopped, and the two phase mixture is magnetically stirred at room temperature for 1.5 hr. The organic layer is evaporated, and the residue is acidified with hydrochloric acid (5%). The crude product is filtered, washed with water, and dried to give a pale-yellow solid, mp 139-142°

(4.61 g). The material, in two approximately equal portions, is purified by chromatography (silica gel column, 2.4 × 30 cm, ethyl acetate) to afford the product as a colorless solid (3.32 g, 69% yield).

4-Ethylsulfonyl-1-naphthalenesulfonamide-^{15}N The above-mentioned product (3.32 g, 12.4 mmol), acetic acid (19 ml), and hydrogen peroxide (30%, 9.5 ml, approx. 90 mmol) are heated on a steam bath for 1.5 hr, diluted with water (125 ml), and cooled in an ice bath. The product is filtered, washed with water, and dried to give pale-yellow needles, mp 198-199° dec. (3.0 g, 81% yield). Recrystallization from ethanol (Norit) gives colorless plates, mp 199-200° (2.5 g).

Notes

1 An alternative scheme in which the label might be introduced in the last step by ammonolysis of 4-ethyl-sulfonyl-1-naphthalenesulfonyl chloride is discussed. The method is satisfactory for preparation of nonisotopic ENS, but when ammonia is not present in excess a dimeric sulfonimide by-product is produced.

References

[1]T. W. Whaley and G. H. Daub, "Syntheses with Stable Isotopes: 4-Ethylsulfonyl-1-naphthalenesulfonamide-^{15}N," *J. Labelled Compds.*, **13**, 481-485 (1977).

POTASSIUM DIHYDROGEN PHOSPHATE-^{18}O$_4$

$$H_2{}^*O \quad \xrightarrow[\text{2. KOH, H}_2\text{O}]{\text{1. PCl}_5} \quad KH_2P^*O_4$$

Procedure[1]

Water-^{18}O (1.0 ml, 50 mmol) is frozen by Dry Ice in a test tube (18 × 150 mm, Note 1) that is closed with a one-hole rubber stopper fitted with a glass Y-tube having one arm connected to a calcium chloride drying tube and the other to a water aspirator. Phosphorus pentachloride (2.90 g, 13.9 mmol) is transferred quickly to the frozen water. Hydrogen chloride from the resulting exothermic reaction is removed by the water aspirator (Note 2). The reaction is brought to room temperature, warmed to 80° over 90 min to remove most of the hydrogen chloride, cooled to room temperature, and treated with potassium hydroxide (2 M) to adjust the pH to 4.66. Ethanol (2 volumes) is added to precipitate the product, which is filtered, washed with ethanol and ether, and dried at 100° for 1 hr to give white crystals (Note 3).

Notes

1 The tube should not be smaller, and a larger scale of reaction is not recommended.

2 Occasionally the reaction must be initiated by partially melting the water.

3 The product is contaminated with a slight amount of potassium chloride. If chloride-free material is desired reprecipitation with 66% ethanol is used. The yield is approx. 85%; ^{31}P-NMR, MS, UV, and chemical analytical data are given.

References

[1]J. M. Risley and R. L. Van Etten, "A Convenient Synthesis of Crystalline Potassium Phosphate-[18]O$_4$ (Monobasic)," *J. Labelled Compds.*, **15**, 533-538 (1978).

METHYLPHOSPHONOUS-^{13}C DICHLORIDE

$$*CH_3MgI \xrightarrow[Et_2O]{(Et_2N)_2PCl} (Et_2N)_2P*CH_3 \xrightarrow{PhPCl_2} *CH_3PCl_2$$

Procedure[1]

Bis(diethylamino) (methyl-^{13}C)phosphine A solution of the Grignard reagent from methyl-^{13}C iodide (2.0 g, 11 mmol) and magnesium (0.34 g, 14 mmol) in ether (10 ml) is filtered through glass wool and added dropwise over 0.5 hr to a stirred and cooled (Dry Ice) solution of bis(diethylamino)chlorophosphine (2.52 g, 12 mmol) (Note 1) in ether (15 ml). The mixture is allowed to warm to room temperature and then refluxed for 0.5 hr (Note 2). The solution is cooled, decanted, and evaporated, and the residue is distilled through a short Vigreux column to give a clear, colorless liquid, bp 85° at 10 mm Hg (1.72 g, 82% yield).

Methylphosphonous-^{13}C dichloride Dichlorophenylphosphine (4.0 g, 22.4 mmol) is added to the product above (1.72 g, 9.1 mmol) (Note 3), a magnetic stirring bar is added, and the flask is fitted for distillation. The mixture is stirred gently and heated in a silicone-fluid bath. At a bath temperature of 200-220° the product distills as a clear, colorless liquid, bp 81° at 750 mm Hg (0.80 g, 76% yield).

Notes

1 Directions are given by Colquhoun and McFarlane for preparing this reagent from diethylamine, phosphorus trichloride, and triethylamine.

2 The solid matter becomes gummy and adheres to the walls of the flask.

3 Heat is evolved.

References

[1]I. J. Colquhoun and W. McFarlane, "The Small-Scale Synthesis of ^{13}C-Labelled Dichloromethylphosphine (Methylphosphonous Dichloride)," *J. Labelled Compds.*, **13**, 535-537 (1977).

DIETHYL (BENZYL-1-^{13}C)PHOSPHONATE

Procedure[1]

(Bromomethyl)benzene-1-^{13}C (1.44 g, 8.4 mmol) is added dropwise over 15 min to triethyl phosphite (1.39 g, 8.5 mmol). The mixture is heated at 150° for 1 hr and distilled under vacuum to give the product, bp 98-100° at 0.5 mm Hg (1.80 g, 94% yield).

References

[1]G. A. Braden and U. Hollstein, "Synthesis of 1-Phenyl-2-phenyl-1-^{13}C-ethene-1-^{13}C (*trans*-Stilbene) and Derivatives," *J. Labelled Compds.*, **12**, 507-516 (1976).

^{13}C-ALDOSES

Procedure[1,2] (Note 1)

A solution of the starting aldose and potassium cyanide-^{13}C (each at a concentration of 0.1 M) (Note 2) is adjusted with acetic acid (2 M) to the optimum pH for the reaction, which is maintained throughout the reaction by addition of acetic acid (2 M) or sodium hydroxide (1 M) as required. When reaction is complete (Notes 3 and 4) the pH is lowered to 4.2 ± 0.1, and hydrogenation is carried out at 60 psi over prereduced palladium-barium sulfate (5%, 62 mg/mmol of nitrile) for 2 hr at 25° with vigorous agitation (Note 5). The solution is filtered, adjusted to pH 2.8 ± 0.2 with Dowex-50(H$^+$), filtered, and concentrated to a syrup, which is dissolved in water and adjusted to pH 9.5 with sodium hydroxide. After 30 min the solution is applied to a column of Dowex-1(OAc$^-$) and eluted with water. The effluent is collected in a flask containing Dowex-50(H$^+$) and contains the unreacted starting aldose and product (yield approx. 55% based on starting aldose) (Note 6).

Notes

1 Tables and discussions are given in the references for the particular reaction conditions to be employed for specific compounds in conjunction with general procedures that are outlined. Serial applications of the procedure allow labeling

in positions other than C-1. The following compounds are listed as being prepared by these procedures: glycolonitrile-1-^{13}C, glycolaldehyde-1-^{13}C, DL-glyceronitrile-1-^{13}C, DL-glyceric-1-^{13}C acid, DL-glyceraldehyde-1-^{13}C, D-erythronitrile-1-^{13}C, sodium D-erythronate-1-^{13}C, D-erythrose-1-^{13}C, D-threonitrile-1-^{13}C, sodium D-threonate-1-^{13}C, D-threose-1-^{13}C, D-lyxose-1-^{13}C, D-xylose-1-^{13}C, D-ribose-1-^{13}C, D-arabinose-1-^{13}C, D-glucose-1-^{13}C, D-mannose-1-^{13}C, D-galactose-1-^{13}C, D-talose-1-^{13}C, D-allose-1-^{13}C, D-altrose-1-^{13}C, D-idose-1-^{13}C, and D-gulose-1-^{13}C; DL-glyceronitrile-2-^{13}C, DL-glyceraldehyde-2-^{13}C, DL-erythronitrile-2-^{13}C, DL-erythrose-2-^{13}C, DL-threonitrile-2-^{13}C, DL-threose-2-^{13}C, D-glucose-2-^{13}C, and D-mannose-2-^{13}C; DL-erythronitrile-3-^{13}C, DL-erythrose-3-^{13}C, DL-threonitrile-3-^{13}C, and DL-threose-3-^{13}C; and DL-erythronitrile-1,2-^{13}C, DL-erythrose-1,2-^{13}C, DL-threonitrile-1,2-^{13}C, and DL-threose-1,2-^{13}C${}_2$.

2 The apparatus consists of a vessel (such as a multineck 25-ml flask) equipped to measure pH in the stirred mixture, to make additions of acid or base, and to remove samples for analysis. Through choice of pH and temperature it has been found to be possible to establish reaction rates for the condensation with cyanide that are convenient to follow with ^{13}C-NMR.[6] For example, conversion of D-erythrose proceeds at a very satisfactory rate at pH 6.5 and 0° to afford the epimers in approx. 1:1 ratio.

3 Samples are withdrawn at intervals, and the extent of reaction is determined by GC (trimethylsilyl derivatives).

4 With 1:1 ratios of cyanide to aldose at least 10% of the starting aldose remains unreacted. A three-fold excess of cyanide gives the aldononitriles in better than 90% (chemical) yields; excess cyanide can be recovered.

5 Reduction times and pH vary, particularly for the short-chain aldononitriles.

6 Total incorporation of cyanide is approx. 85%; that is, 55% in product, 20% in aldonic acids, and 10% in aminoalditols. The mixture is separated and purified by chromatography on ion-exchange resins.[3]

Other Preparations

Photosynthetic incubation of excised tobacco leaves with carbon-^{13}C dioxide incorporates approx. 60% of the isotope (with only slight dilution of isotopic concentration) into ^{13}C-starch (major product), sucrose-^{13}C${}_{12}$, glucose-^{13}C${}_6$, and fructose-^{13}C${}_6$.[4,5] Somewhat different distributions of products are produced by leaves of other plants (Swiss chard, canna).[5] Similarly, various species of the marine red alga *Gigartina* (kelp) give galactosylglycerol-^{13}C${}_9$ (43% isotopic yield containing endogenous, nonisotopic material), which is hydrolyzed to galactose-^{13}C${}_6$ and glycerol-^{13}C${}_3$.[5]

References

[1] A. S. Serianni, H. A. Nunez, and R. Barker, "Carbon-13-Enriched Carbohydrates. Preparation of Aldononitriles and Their Reduction with a Palladium Catalyst," *Carbohydr. Res.*, 72, 71-78 (1979).

[2] A. S. Serianni, E. L. Clark, and R. Barker, "Carbon-13-Enriched Carbohydrates. Preparation of

Erythrose, Threose, Glyceraldehyde, and Glycolaldehyde with ^{13}C-Enrichment in Various Carbon Atoms," *Carbohydr. Res.*, **72**, 79-91 (1979).

[3]O. Samuelson, "Partition Chromatography of Ion-Exchange Resins," *Methods Carbohydr. Chem.*, **6**, 65-75 (1972); J. K. N. Jones and R. A. Wall, "The Separation of Sugars on Ion-Exchange Resins," *Can. J. Chem.*, **38**, 2290-2294 (1960).

[4]V. H. Kollman, J. L. Hanners, J. Y. Hutson, T. W. Whaley, D. G. Ott, and C. T. Gregg, "Large-Scale Photosynthetic Production of Carbon-13 Labeled Sugars: The Tobacco Leaf System," *Biochem. Biophys. Res. Commun.*, **50**, 826-831 (1973).

[5]V. H. Kollman, C. T. Gregg, J. L. Hanners, T. W. Whaley, and D. G. Ott, "Large-Scale Photosynthetic Production of Carbon-13 Labeled Sugars," in *Proceedings of the First International Conference on Stable Isotopes in Chemistry, Biology and Medicine*, Argonne National Laboratory, Argonne, IL, May 9-11, 1973 (P. D. Klein and S. V. Peterson, Eds.), USAEC Report CONF-730525 (1973), pp. 30-40.

[6]R. M. Blazer, Los Alamos Scientific Laboratory, Los Alamos, NM, personal communication, January 1980.

5-ACETAMIDO-3,5-DIDEOXY-D-*glycero*-
D-*galacto*-2-NONULOSONIC-1-^{13}C ACID
(*N*-ACETYLNEURAMINIC-1-^{13}C ACID; ^{13}C-NANA)

Procedure[1]

4-Acetamido-2,4-dideoxy-D-*glycero*-D-*galacto*-octose (0.30 g, 1.1 mmol) (Note 1) in water (2.5 ml) is treated with a solution of sodium cyanide-[13]C (0.15 g, 3.07 mmol) in water (1.7 ml). The mixture is kept at 4° for 7 days, heated to 70° for 3 hr while a slow stream of nitrogen is blown over the surface, cooled, and passed down a column of Rexyn-101(H$^+$) ion-exchange resin (5 ml). The eluate on concentration affords almost equal quantities of 5-acetamido-D-*erythro*-L-*manno*-nononic-1-^{13}C acid and 5-acetamido-D-*erythro*-L-*gluco*-nononic-1-^{13}C acid (0.26 g, 79% yield, Note 2). The mixture of acids (0.25 g) in water (6 ml) is adjusted to pH 8.5 with potassium hydroxide (0.1 *N*) and kept at 20° for 2 hr while the pH is maintained at 8.5. Potassium chlorate (0.12 g, 1 mmol) and the oxidizing catalyst (5.4 ml) prepared by stirring vanadium pentoxide (75 mg) in hydrochloric acid (conc., 4.5 ml) are added, and the mixture is stirred at 20° for 20 hr, diluted with water (15 ml), and extracted with chloroform (15 ml). The water layer is passed down a column of Dowex-50(H$^+$) resin (14 ml) and then a column of Dowex-1-X8(formate) resin (18 ml). The latter column is eluted with formic acid (5%, 50 ml), and the eluate is concentrated to a syrup (200 mg), which is dissolved in water (1 ml) and chromatographed on a column of Dowex-1-X8(formate resin (1 × 24 cm). The column is eluted with a formic acid gradient (0-10%, 300 ml), and the fractions that contain product (by paper chromatography) are combined and concentrated at reduced pressure to give the crystalline product, mp 182-190°, [α]$_D$ -32° (*c* 0.8, water) (128 mg, 51% yield) (Note 3).

Notes

1 Directions are given by Benzing-Nguyen and Perry for preparing the starting material from 2-acetamido-2-deoxy-D-mannose by stepwise extension of the carbon chain with procedures that utilize nitromethane and modified Nef reactions.

2 The yield is based on the starting octose.

3 TLC, GC, and ^{13}C-NMR data are given.

References

[1]L. Benzing-Nguyen and M. B. Perry, "Stepwise Synthesis of N-Acetylneuraminic Acid and N-Acetyl[1-^{13}C] neuraminic Acid," *J. Org. Chem.*, **43**, 551-554 (1978).

2-AMINO-2-DEOXY-D-GLUCOSE-^{15}N HYDROCHLORIDE

Procedure[1]

2-(Benzylamino-^{15}N)-2-deoxy-D-glucononitrile D-Arabinose (1.25 g, 8.3 mmol), benzylamine-^{15}N (approx. 9.8 mmol) (Note 1), and ethanol (4 ml) are refluxed gently with stirring for 15 min. The solution (Note 2) is transferred with ethanol (15 ml) into a large test tube and treated with a steady stream of hydrogen cyanide (Note 3) for 2 hr (Note 4) to afford the crystalline product, mp 131-132° (1.80 g, approx. 69% yield, Note 5).

2-Amino-2-deoxy-D-glucose-^{15}N hydrochloride The foregoing nitrile (1.781 g, 6.7 mmol) in hydrochloric acid (0.5 M, 40 ml) is hydrogenated at ambient temperature and pressure in the presence of palladium chloride (180 mg). Hydrogen consumption is 295 ml after 3 hr, and the slightly yellow solution is filtered and concentrated to give the product, $[\alpha]_D^{22}$ +69.2° (709 mg, 49% yield) (Note 6).

Notes

1 The amine is the total product from reduction (in approx. 80% yield) of benzamide (1.5 g, 12.3 mmol).
2 The slightly yellow solution presumably contains the glycosylamine.
3 Appropriate precautions are observed.
4 The crystalline nitrile begins to form after 1 hr, and the reaction appears complete after 2 hr.
5 The yield is 55% based on benzamide-^{15}N.
6 Descending paper chromatography reveals only one spot.

References

[1]U. Hornemann, "Synthesis of 2-amino-2-deoxy-D-glucose-^{15}N and of 2-amino-2-deoxy-L-glucose-2-^{14}C," *Carbohydr. Res.*, **28**, 171-174 (1973).

D-RIBOSE-2-^{18}O

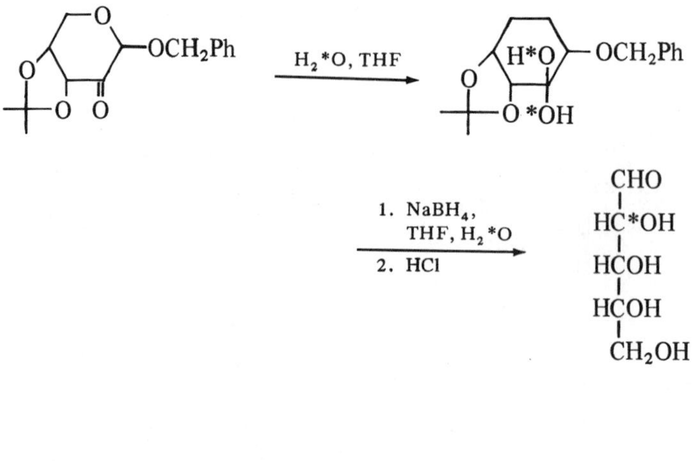

Procedure[1]

A solution of benzyl 3,4-*O*-isopropylidene-β-D-erythropentopyranos-2-uloside (0.45 g, 1.6 mmol) in water-^{18}O (0.200 g, 10.0 mmol, Note 1) and tetrahydrofuran (1.4 ml) is kept at room temperature for 16 hr and then evaporated at reduced pressure (Note 2). The residue (Note 3) is dissolved in tetrahydrofuran (4.0 ml) containing water-^{18}O (0.050 g), treated with sodium borohydride (0.45 g, 12 mmol), stirred for 45 min, treated with water, and extracted with ether (5 × 15 ml). The ether solution is concentrated, treated with petroleum ether, and cooled to give coarse crystals of benzyl 3,4-*O*-isopropylidene-β-D-ribopyranoside-2-^{18}O, mp 92-94° (0.30 g, 66% yield) (Note 4).

The pyranoside (1.85 g, 6.8 mmol) is hydrolyzed in boiling hydrochloric acid (1 *N*, 50 ml) for 75 min. The cooled solution is washed with ether and passed rapidly through a column of Dowex-1X8(OH⁻) (1.5 × 30 cm), and the column is eluted with water until the eluate ceases to react with Fehling's solution (Note 5). The solution is evaporated at reduced pressure to give the colorless product (0.90 g, 88% yield) (Note 6).

Notes

1 The water is 97.9 at. % ^{18}O.

2 The water is recovered quantitatively in a cold trap and is reused in subsequent preparations.

3 Recovery of the mixture of labeled ketone and its hydrate is quantitative.

4 The oxygen at position 2 contains 51.4 at. % excess ^{18}O (average from five preparations). From the mother liquors is obtained the epimer, benzyl 3,4-*O*-isopropylidene-β-D-arabinopyranoside-2-^{18}O, mp 55-58° (18% yield).

5 To avoid epimerization the quantity of resin should not be larger than required for complete neutralization of the acid.

6 The product is converted to 1-*O*-acetyl-2,3,5-tri-*O*-benzoylribose-2-^{18}O, adenosine-2'-^{18}O, adensine-2'-^{18}O 5'-monophosphate, and adenosine-2'-^{18}O 5'-triphosphate.

References

[1]H. Follmann and H. P. C. Hogenkamp, "The Synthesis of Adenine Nucleotides Containing Oxygen-18," *J. Am. Chem. Soc.*, 92, 671-677 (1970).

1,2,3-TRI-*O*-ACETYL-5-*O*-BENZOYL-D-RIBOFURANOSE-3-^{18}O

$$\xrightarrow[\text{EtOH}]{\text{H}_2,\ \text{PtO}_2}$$

BzO—⟨O⟩—O
*OH O—✳ (isopropylidene)

$$\xrightarrow[\text{H}_2\text{SO}_4]{\text{Ac}_2\text{O},\ \text{AcOH}}$$

BzO—⟨O⟩—OAc
Ac*O OAc

Procedure[1]

1,2-O-Isopropylidene-5-O-benzoyl-α-D-*erythro*-pentofuranos-3-ulose (0.25 g, 0.85 mmol) is incubated in tetrahydrofuran (0.6 ml) containing water-^{18}O (0.290 g, 14.5 mmole, Note 1) for 6 hr at room temperature. The solution is evaporated (Note 2) to give a white solid residue, which is recrystallized from ether; mp 90-108° (0.265 g, 100% yield, Note 3). The product from several preparations (1.56 g, 5.3 mmol) in ethanol (40 ml) containing platinum dioxide (0.15 g) is hydrogenated at room temperature and pressure for 20 hr to afford 1,2-O-isopropylidene-5-O-benzoyl-α-D-ribofuranose-3-^{18}O (1.10 g, 70% yield), which is recrystallized from benzene-cyclohexane to give long prisms (Note 4).

To a cooled solution of the latter product (1.20 g, 4.0 mmol) in acetic acid (17 ml) and acetic anhydride (3 ml) is added sulfuric acid (conc., 1.0 ml) dropwise, and the darkening solution is allowed to reach room temperature. After 16 hr it is diluted with ice water and extracted with ether (4 × 30 ml). The extract is washed twice with water and evaporated at reduced pressure, and the residue is coevaporated with alcohol, dissolved in alcohol, decolorized with charcoal, and concentrated to afford crystals, mp 116-118° (0.39 g). From concentration of the mother liquor is obtained additional product (0.22 g, 40% total yield) (Note 5).

Notes

1 The water is 96.5 at. % ^{18}O.

2 The recovered tetrahydrofuran-water-^{18}O mixture is treated with additional water-^{18}O (0.040 g, 2 mmol) and used to incubate another ketone sample.

3 The product is a mixture of the labeled ketone and its hydrate and has an isotopic content of 17.14 at. % ^{18}O.

4 Isotopic content is 9.56 at. % ^{18}O or 56.2% excess in the 3-hydroxyl group.

5 The product is converted to adenosine-3′-^{18}O (45% yield), adenosine-3′-^{18}O 5′-monophosphate, and adenosine-3′-^{18}O 5′-triphosphate.

Other Preparations

D-Ribose-3-^{18}O has been obtained[1] by Fischer-MacDonald degradation of D-allose-

3-^{18}O (through the dithioacetal and disulfone); however, the yield is considered unsatisfactory for labeling, and the method above is preferred.

References

[1] H. Follmann and H. P. C. Hogenkamp, "The Synthesis of Adeline Nucleotides Containing Oxygen-18," *J. Am. Chem. Soc.*, **92**, 671-677 (1970).

D-ALLOSE-3-^{18}O

Procedure[1]

An emulsion of 1,2:5,6-di-*O*-isopropylidene-α-D-*ribo*-hexafuranos-3-ulose (1.2 g, 4.65 mmol) in water-^{18}O (1.0 g, 50 mmol, Note 1) is stirred rapidly for 5-10 min until it solidifies. Evaporation at reduced pressure (Note 2) leaves the hydrate as a white solid (1.29 g, 100% yield), which is dissolved in tetrahydrofuran (10 ml), treated with sodium borohydride (1.9 g, 50 mmol), and stirred at room temperature for 1 hr. Water (50 ml) is added, the product is extracted with ethyl acetate (7 × 15 ml), and the extract is evaporated at reduced pressure. The colorless syrup is crystallized from benzene-petroleum ether to give 1,2:5,6-di-*O*-isopropylidene-α-D-allofuranose-3-^{18}O, mp 73-75° (0.71 g, 59% yield, Note 3) (Note 4).

A solution of the product above (2.20 g, 8.5 mmol) in water-methanol (1:1, 100 ml) is treated with sulfuric acid (0.3 *N*, 100 ml) and refluxed for 2 hr. The mixture is neutralized with barium carbonate, filtered with the aid of water (500 ml), and evaporated at reduced pressure. The colorless syrup solidifies after repeated evaporations with ethanol to afford the product as a dry white powder (1.53 g, 100% yield) (Note 5).

Notes

1 The water is 96.5 at. % ^{18}O.

2 The excess water-^{18}O (0.91 g) is recovered quantitatively.

3 The product is 41.8 at. % excess ^{18}O at position 3; this is smaller than the theoretical concentration (47%), probably because of incomplete hydrate forma-

tion in the heterogeneous medium.

4 From the combined mother liquors of several preparations is obtained 1,2:5,6-di-O-isopropylidene-α-D-glucofuranose-3-^{18}O, mp 106-109° (10% yield).

5 Directions are given for conversion to the diethyl dithioacetal and to 1-deoxy-1,1-bis(ethylsulfonyl)-D-allitol-3-^{18}O (22% yield).

References

[1]H. Follmann and H. P. C. Hogenkamp, "The Synthesis of Adenine Nucleotides Containing Oxygen-18," *J. Am. Chem. Soc.*, **92**, 671-677 (1970).

ADENOSINE-5'-^{18}O

Procedure[1]

Adenosine-5'-aldehyde (50 mg, 0.19 mmol) is stirred with tetrahydrofuran (0.6 ml) and water-^{18}O (0.250 g, 12.5 mmol) for 16 hr at room temperature (Note 1), and the mixture is evaporated at reduced pressure. The residue is suspended in tetrahydrofuran (1 ml) containing sodium borohydride (0.10 g, 2.6 mmol) and water-^{18}O (50 mg, 2.5 mmol), stirred for 1 hr, water (0.5 ml) is added, and stirring is continued for 20 min to give a clear solution. Tetrahydrofuran is removed at reduced pressure, and the aqueous solution is neutralized with acetic acid and concentrated. Salts are removed by paper chromatography (butanol-water) or by passage over Dowex-1X2(OH⁻) (1 × 30 cm) and elution with methanol (60%) to give the nucleoside (34 mg, 68% yield) (Note 2).

Notes

1 The water is 97 at. % ^{18}O. The aldehyde slowly dissolves.

2 Isotopic concentration is 50 at. % ^{18}O at the labeled position.

References

[1]H. Follmann and H. P. C. Hogenkamp, "The Synthesis of Adenine Nucleotides Containing Oxygen-18," *J. Am. Chem. Soc.*, **92**- 671-677 (1970).

17α-(HYDROPEROXY-$^{18}O_2$)PROGESTERONE

1. *O_2, t-BuOK,
   t-BuOH, THF

2. HOAc, MeOH

Procedure[1]

3-Ethoxy-17α-(hydroperoxy-$^{18}O_2$)pregna-3,5-dien-20-one   3-Ethoxypregna-3,5-dien-20-one (1 g, 2.9 mmol), tetrahydrofuran (12 ml), and *tert*-butyl alcohol (4 ml) in a 50-ml flask on a vacuum line are stirred and cooled to $-10°$. Under a nitrogen atmosphere potassium *tert*-butoxide (0.81 g, 7.2 mmol) is added, the reaction vessel is cooled in liquid nitrogen, and the system is degassed and evacuated to less than 0.1 mm Hg. Oxygen-$^{18}O_2$ (100 ml, approx. 4.46 mmol) is admitted, the stopcock joining the reaction flask to the system is closed, and the liquid nitrogen bath is replaced with an ice-salt bath. The mixture is stirred at about $-10°$ for 90 min, cooled in liquid nitrogen, opened to the manifold again, and the process above is repeated. The yellowish reaction mixture is poured into ice-cold sodium chloride solution (saturated) and brought to pH 7. The two-phase mixture is extracted exhaustively with ethyl acetate and ether, and the extracts are washed with water, dried over sodium sulfate, and evaporated at reduced pressure at $45°$ to give the product as a white residue (640 mg, 38% yield, Note 1). Two recrystallizations from dioxane-water give colorless crystals, mp 136-138°.

17α-(Hydroperoxy-$^{18}O_2$)-4-pregnene-3,20-dione   The hydroperoxide above (500 mg, 1.32 mmol) is dissolved in warm methanol, acetic acid (25 ml) is added, and the solution is cooled rapidly to room temperature, stirred for 120 min, concentrated at reduced pressure, and poured into ice-cold water (100 ml). The white precipitate is recrystallized from dioxane-water to give colorless needles, mp 183-185° (420 mg, 91% yield) (Note 2).

Notes

1 The yield based on the organic starting material is 58%.

2 ^{1}H-NMR and MS data are presented and discussed.

References

[1]P. Falardeau and L. Tan, "Labeled Steroids of Potential Biological Interest: Synthesis and Properties of ^{18}O-Labeled 17α-Hydroperoxyprogesterone," *J. Labelled Compds.*, 10, 239-248 (1974).

Index

Page numbers are given for products of synthetic procedures in **bold-face** type, for reactants in ordinary type, and for other citations in *italics*. Isotopic designations are secondary to chemical names in determining placement of entries.

215